Physiological Disorders of Vegetable Crops

The Author

Dr. Kanaya Lal Bhat, Professor (Retired) Division of Vegetable Science and Floriculture, Sher-e-Kashmir University of Agricultural Science and Technology, Jammu (J&K), received his Ph.D. in Vegetable Science from HPKV, Palampur in 1994. Before his superannuation, he was teaching to M.Sc. and Ph.D students at SKUAST –Srinagar and SKUAST –Jammu for about twenty years and has guided a number of students in the discipline of Vegetable production. He has numerous research papers and has written extensively in scientific papers and for the general public. Dr. Bhat has participated in several national seminars and workshops on vegetable production. The author has to his credit three books viz., Vegetables: Untapped Potential (2007), Physiological Disorders of Vegetable Crops (2009) and Brinjal (2011).

Physiological Disorders of Vegetable Crops

By

K.L. Bhat

2024

Daya Publishing House®

A Division of

Astral International (P) Ltd

New Delhi 110 002

First Impression, 2009
Reprinted, 2024

ISBN 978-81-7035-591-5

Publisher's note:

Every possible effort has been made to ensure that the information contained in this book is accurate at the time of going to press, and the publisher and author cannot accept responsibility for any errors or omissions, however caused. No responsibility for loss or damage occasioned to any person acting, or refraining from action, as a result of the material in this publication can be accepted by the editor, the publisher or the author. The Publisher is not associated with any product or vendor mentioned in the book. The contents of this work are intended to further general scientific research, understanding and discussion only. Readers should consult with a specialist where appropriate.

Every effort has been made to trace the owners of copyright material used in this book, if any. The author and the publisher will be grateful for any omission brought to their notice for acknowledgement in the future editions of the book.

Published by : **Daya Publishing House®**
A Division of
Astral International Pvt. Ltd.
– ISO 9001:2015 Certified Company –
4736/23, Ansari Road, Darya Ganj
New Delhi-110 002
Ph. 011-43549197, 23278134
E-mail: info@astralint.com
Website: www.astralint.com

Laser Typesetting : **Classic Computer Services**, Delhi - 110 035

Printed at : **Replika Press Pvt. Ltd.**

— Dedicated to —

The pestilence, forbearance and resoluteness of tiny patriotic community of Kashmir, who inspite of the onslaughts, have come out of it with renewed pride in their spiritual, cultural and educational traditions and the ethos they cherish.

Preface

Vegetables form an important part of our diet. They contain a number of nutritionally important compounds such as vitamins, minerals, carbohydrates and proteins. They often attract the consumer for their aesthetic qualities like flavour, colour and texture and contain a high percentage of water. Consequently, they exhibit relatively high metabolic activity when compared to other plant derived foods as seeds.

Vegetable production increased more rapidly than other crop production. This was due to improved facilities for production, processing and distribution to educational and promotional programmes dealing with the importance of vegetables in diet and to the rising purchasing power, changing food habits and life styles. The quality of vegetables that we consume is influenced by deficiency or excess of mineral elements, hormonal imbalance, improper pollination or fertilization, injuries due to adverse climatic and growing conditions and some physiological factors. Sometimes more than one factor may be responsible for physiological disorder. Almost all major vegetable crops are prone to various types of disorders affecting different plant organs thus rendering them unfit for human consumption, therefore, control of disorder is essential for profitable production of the crop. Physiological disorder is the breakdown of tissue that is not caused by either invasion by pathogen or by mechanical damage but may develop due to adverse environment, especially temperature or to nutritional deficiency during growth and development. These disorders may develop during growth, storage or in transit thus render the produce unmarketable. Examples of physiological problems in vegetable crops are blossom-end rot of fruit (a leather like decay of the blossom-end of several vegetable fruits associated with calcium deficiency). Secondary growth of potatoes and bolting of many biennial and perennial crops (a pre-mature flowering that occurs when the crops are exposed to inappropriate day lengths or temperatures).The production of premature flower stalk (bolting) in spinach and ensuing seed production can render the plant unmarketable. This can occur when spinach is grown under long days and warm conditions. Other disorders that effect the economical production of the crop are sun scald, puffiness, cat face, growth cracks, blotchy ripening (tomato),chlorosis & cracked stem (celery), hollow heart, black heart & greening of potato, russet spotting (lettuce and sweet potato), buttoning, whiptail & browning (cauliflower), splitting & doubling of bulbs in onion, albinism (straw berry), red heart (lettuce), high and low temperature during developmental phase and after harvest also render the produce unfit for market.

Therefore, increased production of vegetables will have significance provided they reach the consumer in good condition, that is without any blemish or misshapen, this can be possible only when they are grown under adequate and balanced nutrition essential for sustaining crop productivity. For this reason, the soil critical level of a nutrient defined for a vegetable may not apply to another crop species.

Therefore, an attempt has been made in this book to provide scientific knowledge on the role of both major and minor nutrients, adverse climatic and other related factors in the growth and development of physiological disorders of vegetable crops, making them unfit for human consumption. In this endeavour many sources such as books, bulletins, scientific papers, journals etc. related with the topic have been consulted, and have tried to give as much information as I could regarding symptoms, causes, and role of elements and temperature associated in the development of these disorders.

The author will welcome suggestions and criticisms on the contents of the book and constructive suggestions if any for further improvement will be included in revised edition.

Kanaya Lal Bhat

Contents

Chapter 1
Introduction

Physiological disorders refer to the breakdown of tissue that is not caused by either invasion by pathogens *i.e.* disease causing organisms or by mechanical damage or may refer to non-parasitic or inanimate disease of fruit and vegetable crops. They may develop in response to an adverse environment, especially temperature, or to a nutritional deficiency during growth and development of vegetables. Adequate supply of nutrients, pollution free soil environment and optimum temperature and moisture favour normal growth of the plants. Any deviation from these conditions results in expression of disorders of various magnitudes. The deficiency or excess of any of the nutrient element, heavy metals, soluble toxic salts in the irrigation water, toxic gaseous pollution in the air, unsuitable prevalent temperature, moisture and soil pH have direct effect on plant growth. A number of vegetable crops are highly sensitive to these adverse conditions and depict varied types of symptoms. However, symptoms associated with physiological disorders are difficult to identify and are occasionally confused with injury caused by insects and diseases. Because many physiological symptoms are crop-specific, information on their identification and potential causes are addressed according to their specific crop characteristics. Examples of physiological problems in vegetable crops (Table 1) are blossom-end rot (a leather like decay of the blossom-end of several vegetable fruits associated with calcium deficiency), secondary growth of many biennial and perennial crops (a premature flowering that occurs when the crops are exposed to inappropriate day lengths or temperatures). Environmental fluctuations at different developmental stages can cause several physiological or morphological defects that affect productivity as it renders the fruit unmarketable.

Thus major factors associated with these disorders are deficiency or excess of mineral elements, hormonal imbalance, improper pollination or fertilization or other stresses. Some times more than one factor may be responsible for the cause and hence seldom called as 'Syndrome'. Some of the physiological disorders and injuries due to adverse climatic and growing conditions of vegetable are furnished in Table 18.

Disorder may be defined as an internal and or internal condition of the fruit while blemish refers to only the surface appearance of the fruit. Hansen (1961) defined physiological disorders in fresh fruit as manifestations of injury resulting from reaction to unfavourable environmental conditions at some period during the growing season. Physiological disorders, *i.e.* disorders in which no organism or virus-like entity has been identified as a cause, rather the disorders have been ascribed to environmental causes, such as temperature, moisture, storage conditions and other stresses.

Table 1: Some Important Disorders of Vegetable Crops

Vegetable Crop	*Physiological Disorder*
Beet root	Brown heart or Crown heart or Heart rot, internal break down
Broccoli	Browning, Whiptail
Carrot	Splitting, Cavity spot, Bitterness
Cauliflower	Hollow stem, Whip tail, Browning, Buttoning, Blindness, Leafy ness, Fuzziness, Riceyness.
Cabbage	Tip burn
Celery	Black heart, Cracked stem, Pencil strip, Freezing injury
French bean	Delayed flowering & pod development, Blossom drop and ovule abortion, Cotyledon cracking, Hypocotyl necrosis
Garlic	Bulb sprouting, Splitting
Lettuce	Tip burn, Russet spotting, Rib disorder, Pink rib, Brown stain
Radish	Brown heart
Turnip	Brown heart, Whip tail
Tomato	Blossom-end rot, puffiness, cat face, sun scald, blotchy ripening,, cracking, watery fruit, Gold specks, green back
Potato	Internal brown spot, Black heart, Hollow heart, Greening, Leaf roll
Colocasia, Elephantfoot	Delayed cooking
Sweetpotato, Yam bean	Growth cracks
Taro	Metsubre
Straw berry	Albinism
Cauliflower	Chlorosis, leafiness
Chicory	Internal browning
Sugar beet	Heart rot

Non-parasitic Disorders

These are caused when the plants normal functioning is disturbed by external, physical, chemical or mechanical agents. These agents are mainly from the two media, in which the plant grows, the air and the soil. Human beings, however are a further cause of disturbance, by using inappropriate cultivation techniques.

Nutritional imbalance or deficiency may lead to morphological and physiological sterility. Resistance to climatic causes of sterility is also linked to good nutritional status. It is hard to say which mineral elements are most important. The most important deficiencies can be due to nitrogen, boron and magnesium. Excess levels of elements may also affect flowering, as plants on the soil that are excessively fertile or that have received too much nitrogen fertilizer end to produce abundant vegetative growth but few flowers.

Chapter 2

Physiological Disorders of Tomato

Tomato is one of three most important horticultural crops. The plant originated from Central America and was an important crop for the ancient Aztecs and Incas. In terms of human health, tomato is a major component of daily needs in many countries and constitutes an important source of minerals, vitamins and antioxidant compounds. Tomato fruit quality for fresh consumption is determined by appearance (colour, size, shape, freedom from physiological disorders and decay), firmness, texture, dry matter and organoleptic (flavour) and nutraceutic (health benefit properties. The organoleptic quality of tomato is mainly attributed to its aroma, volatiles, sugar, acid content, while the nutraceutical quality is defined by its minerals, vitamin, carotenoid and flavonoid content.

The production of high quality fruit is also controlled by climate factors and cultural practices. Climate factors are defined as light intensity, embient temperature (day/night), vapour pressure, deficit (VPD), CO_2 enrichment of the atmosphere concentration. Cultural practices such as growing methods (growing media, planting density, cluster size, deleafing, root system), the irrigation regime, and the composition and concentration of nutrient solution, as well as the time of harvest and storage conditions, all influence the quality of the final product. The control of disorders prevalent in vegetable crops is essential for economic production of the crop. These physiological disorders are described in Table 2.

Table 2

Disorder	*Symptoms*
Blossom-end rot	Dark coloured area becomes sunken and leathery
Sun scald	Yellowish blotches appear on leaves
Puffiness	Fruit light weight, lack firmness and partially filled.
Cat face	Blossom-end of ovary turn dark to form a leathery blotch
Growth cracks	Crack occur at the stem end
Blotchy ripening	A diffuse area of yellow to grey colour of the ripe fruit
Watery fruit	Appearance of watery tomato due to an elevated root pressure
Gold specks	Cells containing a granular mass of tiny calcium oxalate crystals
Sun scald	Yellow or white patches on green and ripened fruits
Internal browning	Irregular blanched areas on the side and near stem end
Green back	Stem end portion turns green
Stem abnormality	Slanky growth and hollow stem
Cold injury	Whitish yellow colour later becomes shrinkled
Ghost spot	Similar to Grey mold

Blossom-end rot

The tomato is affected by a serious disorder characterized by a dry rot of the blossom-ends of either green or maturing fruits. Earlier it was described as the rot, black rot, fruit rot, point rot, dry rot or dry-weather rot, but the name blossom-end rot in use is of recent origin, since it gives the best expression of the nature of the trouble.

This is a common and destructive non-parasitic disease that causes extensive yield losses in tomato fruits when they are about half-grown. It consists of a brown discolouration that enlarges until it covers one-third or more of the fruit. It develops as a visible external depression of black necrotic tissue affecting the distal part of the placenta and the adjacent locular contents as well as the pericarp (Willumsen, *et al.*, 1996). The dark tissues shrink and form a flattened or slightly concave area with a dry, leathery surface. There is no rot of the fruit unless the dead tissue is later invaded by fungi or bacteria. Blossom-end rot usually appears when the plants that have made a rapid growth are subject to high temperatures and drought. Then the cells at the blossom-end of the fruits apparently fail to receive enough moisture to maintain their growth, and the tissues break down. Plants that have had heavy applications of nitrogenous fertilizer are especially susceptible to blossom-end-rot. Marked fluctuations in soil moisture increase the likelihood of injury. Plants that are pruned and staked consistently show more blossom-end rot than those growing naturally in otherwise similar conditions. That may be due largely to the restriction of the root system that results from drastic pruning of the tops. The disorder is often prevalent in commercial as well as kitchen garden tomatoes and severe losses may occur if preventive control measures are not taken. Tomato fruits are more sensitive to BER 7 –10 days to 21 days after anthesis. Internal BER, called " black seed" may also be present on the same fruit. In this case, black necrotic tissue is restricted to the adjacent parenchyma tissue around the young seeds and the distal part of the placenta (Adams and Ho,1992). The blossom-end rot of the tomato is a non-infectious disease due to a physiological derangement which results in the death of the protoplasm of the cells of the localized areas within the growing fruits, followed by collapse, drying and discolourations to produce a condition quite properly designed as dry rot.

Symptoms

Blossom-end rot is a calcium related physiological disorder of tomato, the incidence of which is affected by many factors including humidity, root health, and solution pH. Symptoms may appear at any stage in the development of the fruit, but, most commonly are first seen when the fruit is one-third to one-half full size. As the name of the disease implies, symptoms appear only at the blossom-end of the fruit. Spurr in 1959 described blossom-end rot disorder as brown proteinaceous inclusion occurring in the epidermis and pericarp at the stylar end of the fruit. The cell membranes become disorganized and tissue necrosis develops underneath the skin. Lesions develop at the blossom-end of the fruit while still green or ripening on the vine. Initially a small, water soaked spots appear at the point of attachment of the senescent petals, which enlarge rapidly until 1cm or more in diameter and darkens rapidly as the fruit develops. With the result the disease portion shrinks, thus depressing the surface of the lesion, which eventually appears black. The spot may enlarge until it covers as much as one third to one- half of the entire fruit surface, or the spot may remain small and superficial. Large lesions soon dry out and become flattened, black and leathery in appearance and texture. The dead and diseased tissues are then invaded by the secondary organisms. The infested area becomes sunken, leathery and dark coloured. The disease does not spread from plant to plant in the field, nor from fruit to fruit in transit. Since it is a physiological nature, fungicides and insecticides are useless as control measures. The affected fruits are totally unfit for human consumption.

Causes

It is envisaged that it is the interaction between daily irradiance, air temperature, water availability, salinity, nutrient ratios in the rhizosphere, root temperature, air humidity, and xylem tissue development in the fruit, control the incidence of blossom-end rot. For example, Ho *et al.* (1993) found a linear relation between the incidence of BER in all trusses and the product of average daily irradiance and daily temperature throughout the year. It is generally thought that, lack of co-ordination between accelerated cell enlargement, due to high import of assimilates, and the inadequate supply of calcium, due to poor development of xylem within the fruit is linked to fruit susceptibility to this disorder (Ho, *et al.*, 1993; Ho *et al.*, 1995 and Belda, *et al.*, 1996).

Bar Tal and Pressman (1966) reported that it is positively correlated to the leaf K: Ca ratio, but is weekly correlated to the K: Ca ratio in mature fruits. Adam and Ho in 1993 were of the view that a high concentration of organic acids in the fruit may reduce the availability of Ca in the tissues, thus making the fruit more susceptible to this disorder. Willumsen *et al.* (1996) suggested that higher the activity ratios, the higher the risk of blossom-end rot due to a lower up-take of Ca and a reduced availability of Ca in the fruit tissue caused by increased concentration of organic acids in the fruit juices. These authors further reported that by maintaining the ion activity ratios at optimum levels *i.e.* about 0.1 for the first ratio and 0.2 to 0.4 for the second, it is possible to reduce the occurrence of blossom-end rot when salinity of the root zone is increased to improve the fruit taste. While Nonami *et al.* (1995) have concluded that calcium content in fruit is not directly related to blossom-end rot. They therefore, proposed that this physiological disorder results from the expression of some genes under conditions of stress.

The occurrence of this disease is also dependent upon the supply of water and calcium in the developing fruits. Factors that influence the uptake of water and calcium by the plant have an effect on the incidence and severity of blossom-end rot. The disease is especially prevalent when rapidly growing, succulent plants are exposed suddenly to a period of drought. When the roots fail to obtain sufficient water and calcium to be transported up to the rapidly developing fruits, the latter become rotted on their ends. Another common predisposing factor is cultivation too close to the plant; this practice destroys valuable roots, which take up water and minerals. Tomatoes planted in cold, heavy soils often have poorly developed root systems. Since they are unable to supply adequate amounts of water and nutrients to plants during times of stress, blossoms-end rot may result. Soils that contain excessive amounts of soluble salts may predispose tomatoes to the disease, for the availability of calcium to the plants decreases rapidly as total salts in the soil increase. Blossom-end rot is a disorder whose dependence on calcium concentration in the fruits is well known. High K/Ca ratios in the soil may decrease calcium uptake by plants and thereby increase the proportion of fruits showing the blossom-end rot. Although the importance of calcium to fruit is dominated by its uptake by roots and transport (Abou Haadid and Jones,1996). The causal factor for blossom-end rot could be calcium deficiency and water stress may aggravate the symptoms (Spurr,1959). Conclusions from the research of Richardson (1982) reveal that possible increase in blossom-end rot of tomato because of high ammonium levels are inconclusive. It is advisable to minimize the risk by either omitting ammonium salts altogether or apply no more than 10 per cent of the nitrogen input in the ammonium form.

Tomatoes usually grown in soils at relatively low moisture level, get adjusted to such environment, ordinarily do not develop blossom-end rot. On the other hand plants grown at adequate soil moisture level, produce rapid succulent growth, with high rate of transpiration are more subjected to this malady. This is often the result of reduced soil moisture or irregular fluctuation in the moisture level. Therefore, light soils are more inclined to moisture fluctuations, thus more prone to blossom-end rot. It is not that all fruits develop this malady, it is only few fruits may succumb, and the subsequent fruits may develop normally. Horsfall in 1948 was of the view that excessive soil moisture reduced root growth, while as sudden windy hot spell aggravated the disease and correlated the disease incidence to degree of exposure to wind. Foster (1937) reported that increasing amounts of nitrogen were apparently conducive to the occurrence of the disease, provided other conditions are favourable, while as increasing amounts of phosphorus reduce it. Robbins while studying the effect of nutrient solutions of different concentrations, found that at lowest concentrations, growth was highest, but there was no trace of disease, while at the highest concentrations, where there was widest fluctuation in rate of transpiration, 80 per cent fruits became diseased. Schroeder in 1949, reported that transpiration influences blossom-end rot, as he demonstrated it by spraying with a substance that lowered the water loss. With increase in toxicity of chloride and sulphate salts in nutrient solutions, there was increase incidence of the disease due to accumulation of the calcium and magnesium. (Eaton,1942). While Raleigh and Chucka (1944) found the disease to be induced when the nutrient solution was relatively high in nitrogen, sulphur, magnesium, potassium or chlorine and when it was low in calcium. It was further noticed that whenever the calcium content of the fruit fell below 0.20 per cent, the disease occurred on some fruits. Lyon *et al.* (1942) were of the view that incidence of blossom-end rot increased as calcium decreased. Dekock *et al.* (1982) found the influence of truss size on blossom-end rot; thinning increased the incidence of the disorder.

Blossom-end rot is caused by temperature extremes. Fruit set occurs only when night temperatures are between 12.76°C and 23°C. When fruits are not set, blossom fall off.

It was further observed that tomato varieties differ in their susceptibility to blossom-end rot, but variety reaction differs widely. In as much as the disease is the result of a transient physiological abnormality and since varieties differ widely in their rate of fruit set and maturation, a given variety may be very susceptible to the disease when the environment happens to favour it in one situation and may be in a more resistant stage at the critical time in another situation. Blossom-end rot is a disorder whose dependence on calcium concentration in the fruits is well known (Wiersum,1966, Winsor and Adams, 1987, Ho, *et al.*, 1993). High K/Ca ratio's in the soil may decrease calcium uptake by plants and there by increase the proportion of fruits showing blossom-end rot. Although the importance of calcium to fruit is dominated by its uptake by roots and transport (Boom,1973, Dibb and Thompson, 1985). According to Shear (1975) this disorder is considered to be related to localized deficiency of calcium.

Blossom-end rot is usually more severe on pruned and trained plants than on the unpruned and untrained ones. The fact that blossom-end rot is severe during dry weather and is worse on pruned and trained plants than on unpruned ones indicates that the pruning and training allow more moisture to escape from the surface of the soil or else the pruned plants can not get the moisture so well as the unpruned plants. Thompson (1927) indicated that both these factors probably are involved. Experiments on moisture determinations have indicated that greater loss of water from soils growing pruned plants than from the soil growing unpruned ones. The transpiration rate is probably much higher from the pruned plants. The unpruned and untrained plants act as a mulch on the soil, restricting the movement of air and shading the surface so that less moisture is lost by evaporation. In nutshell the cause of this may (*i*) Use of ammonium sulphate for nitrogen supply

(Filipppov, 1961). (*ii*) An imbalance of magnesium and potassium (Taverna,1968), under saline conditions, magnesium is distinctly higher than calcium and thus may cause the disorder (Pasture, 1971). (*iii*) Depletion of calcium in the blossom-end portion (Ward, 1973).

Control

Control of blossom-end rot is dependent upon maintaining adequate supplies of moisture and calcium to the developing fruits. Tomatoes should not be excessively hardened not too succulent when set in the field. They should be planted in well drained, adequately aerated soils. Tomatoes planted early in cold soil are likely to develop blossom-end rot on the first fruits, with the severity of the disease often subsiding on fruits set later. Thus, planting tomatoes in warmer soils helps to alleviate the problem. Irrigation must be sufficient to maintain a steady even growth rate of the plants. Mulching of the soil is often helpful in maintaining adequate supplies of soil water in times of moisture stress. When cultivation is necessary, it should not be too near the plants nor too deep, so that valuable feeder roots remain uninjured and viable. In kitchen gardens, shading the plants is often helpful when hot, dry winds are blowing, and soil moisture is low.

However, there are several ways of reducing the amount of blossom-end rot in tomato fruit.

By maintaining proper calcium levels in tomato fruit. Therefore, the soil for growing tomato should be properly limed, to raise the soil pH to a suitable range for optimum calcium uptake. Calcium must dissolve in soil moisture to be assimilated into the plants.

Single foliar spray of 0.5 per cent calcium chloride solution at the time of fruit development was effective in controlling blossom-end rot (Bhandari, 1963). This disorder can be readily eliminated by the application of calcium salts as pre-harvest spray. It can also be controlled to great extent by balance irrigation and staking Raychoudhury and Lee (1966). Tomato will be less susceptible to blossom-end rot, if they are not pruned too heavily and or if they are not fertilized too heavily with ammonium nitrate. Use of fertilizer low in nitrogen, but high in superphosphate, such as 4-12-4 or 5-20-5, will do much to alleviate the problem of blossom-end rot.

In order to avoid blossom-end rot, different strategies can be used:

1. Choose resistant cultivars
2. Optimize calcium and phosphate supply
3. Keep a dynamic balance between calcium and potassium and between nitrate and ammonium that will ensure sufficient calcium up-take
4. Use low EC
5. Optimize the irrigation frequency in order to avoid water stress or water-lodging
6. Avoid high root temperature (>26° C, Adams and Ho,1993) and low oxygen concentration which will reduce calcium uptake.
7. Avoid excessive canopy transpiration by de-leafing, shading, roof sprinkling, and greenhouse fogging
8. Keep a proper fruit: leaf ratio that will provide an adequate fruit growth rate; and

9. Spray young expanding fruit with 0.5 to 0.65 per cent calcium chloride solution (Igbokwe *et al.*, 1987, Ho, 1999).
10. Transplanting in early April instead of an early June reduces the disorder by about 60 per cent and increases yield by about 20 per cent.

Blotchy Ripening

Blotching ripening or green back is one of the major physiological disorders in tomato fruit. The affected fruits show areas of yellow or orange discolouration on the surface intermixed with normal fruit colour. The blemish is brought about by irregular ripening, particularly on the cheek of the fruit. It is a diffuse area of yellow to grey in otherwise normal red colour. Internally the parenchyma surrounding the vascular bundles of the outer fruit walls becomes necrotic and disorganized, apparently from stress a reaction (Sadik and Minges 1966).The affected tissues may either be opaque or brownish in colour and are lignified or starchy. Hobson (1967) observed that some biochemical features of the green mature fruits are retained in the blotchy tissue during ripening. Blotchy ripening can be de-induced by pre-harvest nutritional status of the crop. Potassium deficiency or excess nitrogen may bring about more blotchy fruits (Winsor *et al.*, 1961). Potassium is reported to increase fruit size and rectifies many disorders like blotchy ripening, vascular browning, white wall and grey wall (Hayslip and Lley, 1967). While Ozbun *et al.* (1967) showed that blotchy ripening was caused by low doses of K. Gallagher (1972) reported that increased level of potassium helps in decreasing the incidence of irregular shaped fruits. According to Trudel and Ozbun (1970) K plays an important role in the process of tomato fruit pigmentation; it increases carotenoids particularly lycopene and decreases chlorophyll. It also exerts a strong influence on acid metabolism in tomato fruits.

Causes

Blotchy ripening or the uneven development of colour on the fruit may be due to temperatures below 15.54°C, root stress from compact or spongy soil, or low levels of potassium in the soil. The discolouration may be random over the surface or confined to the fruit shoulder. Internally, a whitish discolouration of the pericarp and placenta tissue may appear, and in severe cases, brown lignified strands occur in the outer pericarp. Seaton and Gray (1936) attributed this disorder to water deficiency during the periods of excessive transpiration, but the work conducted in England does not support that view. Another view point is that there is definite negative correlation between light intensity *i.e.* sun shine and the incidence of blotchy ripening. As there is decrease in normal starch during the ripening process in the blotchy ripening areas. It is believed that the metabolic changes which bring on blotchy ripening in potassium- deficient plants and to a less degree in nitrogen deficient plants are counteracted by increase in light. In blotchy areas the cell walls of the phloem are thicker than normal. This indicates that disturbed translocation is a factor, and there is also the possibility that carbohydrates being used for cellulose are withheld from the normal ripening processes. Therefore, no single agent can be solely responsible; improper fertilization, especially potassium, soil moisture levels, cultivar and virus disease have been associated with this disorder. It can be minimized by (i) maintaining a balance between nitrogen and potassium fertilizer in the soil. When the level of potassium is very high, the application of magnesium reduces the incidence (Woods,1964) and (ii) by using resistant varieties which have the capacity to utilize potassium more effectively.

Cat Face

Is a scarring, puckering and malformation at the blossom-end of the fruit. It resembles blossom-end rot but is distinct from it. Such fruits have ridges, furrows, indentations and blotches. Abnormal growing conditions during formation of blossom appear to cause distortion of the growth of the cells of the pistil. As a result the cells in the blossom-end of the ovary die and turn dark to form a leathery blotch at the end of the fruit without the progression of symptoms characteristics of blossom-end rot. The amount of malformation is greater than with the latter disease and little or no invasion by secondary organisms follows. The fruits are necessarily discarded or placed in the cull class. This defect may be only a small of corky tissue extending up the sides of fruit. The severe deformation is termed cat facing, while as lesser damage is called blossom scar. The severe scaring usually is accompanied by a highly lobed and asymmetrical shape. Some cultivars are more susceptible than others.

Causes

There are several reasons, which cause cat-facing. (*i*) Cold temperature during flowering and fruit set, and other stress conditions, including excessive heat, 2,4-D contamination, pruning and erratic moisture, have been associated with cat-facing or scarring of the blossom-end of the fruit. The cool and cloudy weather at blossoming time may cause the blossom to stick to the young, developing fruit, resulting in the malformation. (*ii*) Faulty pollination and fertilization due to low temperature cause cat-facing in tomato fruits. (*iii*) The time of nitrogen application may be another probable cause, by altering the transition of vegetative growth into reproductive phase. Delayed pruning, balancing the internal nutrient, regulating temperature, the assimilation rate and the endogenous growth regulators can control cat-facing.

Growth Cracks

Cracks are characterized by rupturing or cracking of the surface of the fruit. It is common wherever tomatoes are grown and often results in large losses. The cracks occur at the end, radiating out from the attachment to the pedicel or being arranged on the shoulder of the fruit in a concentric order with reference to the point of attachment. In addition to these, circular cracking is also found on the skin of fruits. They vary in depth from very shallow microscopic ones to deep breaks in the tissue. Growth cracks affect the appearance of the fruit and become sites for the development of infection. They are common points of entry by fruit-rotting pathogens that cause decay. Varieties vary in their tendency to crack, but most are more or less subject to the trouble. Early defoliation by foliage-blight organisms enhances cracking, and there is tendency for it to increase in warm rainy weather when rapid growth is taking place.

Cracking has been defined as the physical failure of the fruit skin (Milad and Shackel, 1992), and is generally believed to result from stresses activating on the skin. It could be due to normal processes of growth or damage-induced (Walter, 1967).

It is commonly believed that cracking and splitting in a wide range of fruits such as cherry and tomato (Verner and Blodgett, 1931), occurs as result of a sudden increase in the water content of the soil, atmospheric humidity or temperature. Peet (1992) has summarized, the factors contributing to fruit cracking in tomato and concluded that it occurs when there is a rapid net influx of water and solutes into the fruit at the same time that ripening and/or other factors are reducing the strength and elasticity of the fruit skin. After 10 years of study on tomato fruit cracking, Frazier (1947), concluded that fruit cracked most severely after heavy irrigation at the end of a prolonged dry period. Cracking was less severe in plots with frequent irrigation, which prevented excessive drying of the soil, and it was least severe in plots where the soil moisture content remained low throughout the growing season. Shaded fruits cracked much less than those exposed to sun. Environmental factors associated with fruit cracking include soil moisture, rainfall, relative humidity, temperature and exposure to sun light.

There are several types of fruit cracking injury: fruit bursting, radial cracking (star-shaped originating from the peduncle), concentric cracking (circular cracks originating from the peduncle), and cuticle cracking (russeting). Cuticle cracking is among the most commonly observed green house fruit cracking, while radial cracking may also be present. Radial cracking is more likely to develop in full ripe fruits than on mature green and turning stages of maturity. On the other hand, concentric cracking is relatively low on ripe fruits when it may be high on mature green fruits. Fruits exposed to the sun develop more concentric cracking than those that are well covered with foliage, hence pruned, stalked plants are more susceptible to this type of cracking than those allowed to grow without pruning and staking.

The importance of this disorder varies according to: (1) time of year (more severe during spring, summer and the early part of the fall) (Loske *et al.*, 1980); (2) cultivar (Koske *et al.*, 1980); and (3) ambient conditions (Abbott *et al.*, 1986). Fruit cracking is generally associated with rapid movement of water and sugar towards the fruit when cuticle elasticity and resistance are week. This physiological disorder occurs generally six to seven weeks after fruit set (Bakker,1988). The first occurrence of cuticle cracking in cherry tomato fruit, however, was recognized two weeks after anthesis (Ohta *et al.*, 1995). A physical model, based on the theory of shells (Timoshenko and Wionowsky- Krieger,1959) and the linear theory of elasticity (Considine *et al.*, 1974; Considine and Brown,1981), showed that the site of initiation of cuticle cracking is between locular contents, and the most important stress zone is located near the calyx.

Characteristics associated with tomato fruit cracking are (1) fruit shape and size (Gill and Nandpuri,1970); (2) extensible and thick cuticle (Gill and Nandpuri,1970); (3) deep penetration of cutin inside the cuticle (Hankinson and Rao,1979); (4) thick pericarp (Peet,1992); (5) number of fruits per plant (Peet,1992); (6) fruit position in the plant (Peet andWillits,1995); (6) soluble sugar content (Peet,1992); (8) development of vascular tissues in fruit (Cotner *et al.*, 1969); and (9) plant architecture (Peet,1992). Considine and Brown in 1981 suggested that as the fruit increases in size, more physical stress is applied against the epidermis, and this leads to an increasing susceptibility to fruit cracking. While Ehret *et al.*, in 1993 have shown that there was no correlation between fruit size and their relative growth rate, or with fruit susceptibility to cuticle cracking. Small cluster size (one or two fruit per cluster), however, may increase the percentage fruit surface are cracked and the severity of cuticle cracking (Ehret *et al.,* 1983).

The number of fruits per plant (Peet, 1992) and the position on the plant (Peet and Willits,1995) are also very important factors. As a high number of fruits per plant increases the competition between fruit for carbohydrates, thus reducing the supply of sugars and water to each fruit and as a consequence, their susceptibility to radial and cuticle cracking. A fruit leaf ratio of 1.24: 1 to 1.28: 1 is generally optimal. In addition to increase in fruit cracking incidence on the upper clusters (linear function), the percentage of fruit affected by cracking increased from 2 per cent (1st cluster) to 38 per cent, 41 per cent and 45 per cent for clusters 5, 6, and 7, respectively (Peet and Willits 1995). The susceptibility of fruits of upper clusters to cracking can be explained as a higher irradiance received by fruits of upper clusters and a higher fruit temperature. These factors tend to favour pulp expansion towards the interior of the fruit and consequently a weakening of the cuticle.

Generally fruits having a high content of soluble sugars (low solute potential), have a greater supply of water, thereby increase the pressure against the cuticle, thus the susceptibility to cracking (Bussieres, 1995). An imbalance between the supply (influx) and the loss (efflux) of water will also cause fruit cracking. According to Cotner *et al.* (1969) tomato cultivars with highly developed system of vascular tissues are therefore more resistant to this physiological disorder. Peet in 1992 was of the view that plant architecture and planting density play an important role in affecting fruit susceptibility to cracking by modifying the degree of leaf shade on fruit.

During rapid plant growth periods *(e.g.* under high irradiance, PPF), and accelerated cellular enlargement and fruit development require an additional supply of nutrients such as calcium, an important nutrient in the prevention of fruit physiological disorders such as fruit cracking (Simon,1978).

Gibberellic acid being a phytohormone, responsible for fruit ripening and softening of the cuticle, can alter the calcium dynamics at the level of the pericarp (Bush *et al.*, 198) by increasing the elasticity of he cuticle (Larson *et al.*, 1983). Peet (1992) and Larson *et al.* (1983) were of the view that application of gibberellin and calcium reduced the tendency of fruits to cracking.

Causes

Cracking is more common during rainy season, especially when the prevailing temperature is high (above 32.19° C) and it rains after a long dry spell. Whereas the radial cracking which is more common and damaging, is likely to develop in full ripe fruits than in mature green or turning stage maturity. The concentric cracking is relatively low on ripe fruits than mature green and has been reported in fruits exposed to direct sun rays than on fruits covered with foliage. Frazier (1935) suggested that abundant rainfall and high temperatures favour rapid growth and predispose the fruits to growth cracks. The specific cause of rupturing could be uncoordinate tissue expansion during growth or simply a turgidity phenomena. Burst cracking often occurs after harvest and usually is the result of handling. Such cracks generally are exacerbated by erratic moisture conditions. Such affected tomatoes become unmarketable. Thomas (1948) indicated that water on the surface of the fruit is more conducive to cracking than high soil moisture. According to Reynard (1951) the severity of radial cracking of susceptible strains is more closely associated with the number of days that rain fell just preceding picking than with the amount of rain in the same period. Pasture (1971) founded that fruit cracking is due to calcium deficiency.

Although this physiological disorder causes considerable economic losses in greenhouse and field grown tomatoes, where as green house fruit is more vulnerable to fruit cracking losses because most of the cultivars used lack crack resistance and because fruit is generally harvested at the "pink" stage (30–60 per cent of the surface shows pink or red colour) or later (Peet and Willits,1995). Depending on the extent of this physiological disorder, fruit cracking: (1) reduces fruit appeal (Peet and Willits 1995), (2) reduces fruit shelf-life (Hayman,1987); (3) increases fruit susceptibility to pathogens (Hassan, 1978); (4) reduces fruit marketability (Peet, 1992).

Control

Use resistant cultivars like Sioux, Manalucie crack proof and Punjab Chhuhara.

Picking of the fruits before the full ripe stage reduces the incidence of radical cracking.

Soil application of borax at the rate of 10-15 kg/ha or its spray at 0.25 per cent at the fruiting stage reduces the malady.

Bauerle and Short (1977) observed that misting can increase the average fruit size and reduce the incidence of fruit cracking without effecting the yield.

Puffiness

It is also known as hollowness or boxiness, refers to the existence of open cavities between the outer walls and the locular contents in one or more locules (Grierson and Kader, 1986). Puffiness is not a disease, but the term refers to hollow spaces within tomato, between the jelly like plancental matrix and the outer wall and affected fruits tend to be some what angular in form. Puffiness is also referred to as puffy tomato and as pockets. When the fruits are about to reach the normal size, the outer wall of the fruit continues to develop normally, where as the growth of the internal tissues is retarded. As a result fruits becomes light weight, lack firmness and are partially filled. In other words, fruit develop locules that do not fill with gel, and such puffy fruit are very soft. The fruits may show flat surfaces and not be firm; thus they are known as "puffy". This may be due to non-fertilization of the ovule, embryo abortion after normal fertilization and necrosis of vascular and the placental tissue after the fruits are well developed. The affected fruit is hollow and light in weight. The puffy tomatoes are downgrade or are rendered unmarketable in serious cases. They do not travel well because they are soft and are not appreciated by the consumers because of the lack of gel in the locules. The percentage of fruit affected is related to genotype and growing conditions that cause improper pollination (Grierson and Kader,1986).

Causes

This malady is attributed to high or low temperatures,

soil saturation or super-saturation and low soil moisture and other conditions which impede growth through disturbance of normal metabolism (Yarnell *et al.*, 1937). The high temperature and high soil moisture are the main factors responsible occurrence of this non parasitic–disorder. The other important cause is the poor pollination, resulting in poor development of seed bearing tissue and off–shaped fruits. It also may occurs when nitrogen is excessive or when plants are field treated with ripening hormones. Its causes are not fully understood, but heredity or variety is a factor, as is poor pollination. It seems more common in round than in flat varieties and is more prevalent in fruits that are picked green. There is evidence that high temperature favour puffing, especially, if they occur early in the development of the fruit. Fertilizer treatments have been found helpful in some situations. Puffy fruits are light in weight, they lack firmness and partially filled. The casual factors are high or low temperature and low soil moisture. High temperature and high soil moisture are predisposing factors for puffiness of the fruit. It is controlled by more than one gene (Palevitch and Kedar,1963).

Sun Scald

Tomato fruits either green or ripening if exposed to scorching sun, develops sunscald as the exposed tissue bleaches white, becomes wrinkled, and often shows secondary infection. The tissues of the effected fruit has blistered water-socked appearance and sunken areas develop due to rapid desiccation, which may be of white or grey colour in green fruits and yellowish in red fruits. The cultivars with sufficient foliage can overcome this disorder. It may appear on foliage and fruits. The seedlings which are forced under glass, when exposed to bright sun and drying wind scalds. Sometimes in midseason, yellowish blotches appear on leaves, when it rains, and cloudy weather follows abruptly after a long dry spell. The blotchy area dry up rapidly and the tissues become tan or brown and brittle. The green fruits are susceptible to sunscald. The exposed side of the fruit may become yellow and ripen unevenly or the injured area may become white and blister-like. The tissue then loses water rapidly, shrinks and flattens into a greyish, sunken, papery–like lesion. The patches affected remain yellow when the fruit ripens, but frequently the tissues are so severely damaged that patches shrink and the surface dries out. The injured fruits if picked green – wrap trade may show no injury at harvest, but the lesions develop later in storage and transit. The sunscald fruits usually develop secondary fungi and become unfit for consumption.

Causes

Mostly varieties with sparse foliage or where it is defoliated by early blight or Septoria leaf spot, the green fruits scald rapidly on bright–sunny days. Sun scald first appears. Poor foliage cover allows exposure to sun on pruned, staked, sprawling, or un-healthy plants. A high temperature treatment along with a high light intensity may result in 100 per cent sunscald.

Control

Protect the plants from defoliation by disease and insects.

Use varieties with ample foliage

Follow good cultural practices.

Yellow Top

It affects green–shouldered cultivars and may appear after extended period of hot or cold temperatures or after defoliation by infectious diseases. It appears as a patchy area of poor colour on fruit shoulders.

Large Core, Green Gel

This disorder been attributed to extended high or low temperature exposure, particularly with excessively high nitrogen fertilization. Cultivars differ in showing such defects.

Sunburn

Fruits when exposed to direct sun, the exposed tissue bleaches white, thus becomes wrinkled, followed by secondary infection by fruit rot organisms. Cultivars with profuse foliage seldom show this disorder.

Cloud Spot

It consists of irregular, blotchy, white to yellow spots beneath the skin. It is brought about by punctures by stinkbugs and appears on green and ripe fruit, being more conspicuous on the latter.

Pox

It mostly occurs on out door tomatoes, as small dark green specks up to 3mm in diameter are scattered over the surface, coalescing and becoming sunken to form shallow, dark-coloured pits.

Blossom Drop

Blossom drop is caused by temperature extremes. Fruit set occurs only when night temperatures are between 12.76°C and 23.86°C when fruits are not set, blossom fall off.

Waxy Blister or Fruit Tumor

This disorder results in wax like, irregular tumor on the fruit surface, which becomes brown, depressed and cracks as the fruit matures (Treshow,1955). Such blisters may be induced by rubbing green tomatoes and storing at 21°C- 35°C. According to Treshow (1955), rubbing induces the synthesis of more growth regulators with kininlike activity, causing increased cell division and tumorus growth.

Graywall

The symptoms are greyish-brown discolouration which are seen through the outer fruit wall that may appear on small, immature, green fruit, as well as on the ripe fruit. The areas may become slightly depressed and rough and severe browning may occur internally. According to Jenkins *et al.*, 1962, bronzing of green tomato fruit is associated with TMV. Where as Holmes in 1949 described internal browning, which is characterized by death of tissues within the tomato fruit. While Young (1946) has described it as internal browning or core rot. In case of gray-wall, usually there are irregular blanched areas, mostly on the side near the stem end. The whitish areas mostly on the side of the discoloured tissue are visible.

Green Shoulder

There is development of a green-yellowish colour on the fruit shoulder and the fruit wall may be coarse in texture.

Gold Specks

It has been identified as cells containing a granular mass of tiny calcium oxide crystals (Den Outer and Van Veenendaal,1988). These tiny yellowish spots are regular, less than 0.1μm across and often observed around the calyx and shoulders (De Kreij *et al.*, 1992). Their presence affects the external appearance of the fruit (Groossens, 1988) and reduces their shelf-life.

Green Back

The main characteristic of green back is that the stem-end portion turns green. Temperature is the main cause of the disorder, which is increased by defoliation. Ventor (1970) has associated green back with temperature and chlorophyll content of the fruits. At high temperature, ripening inhibited and green back is expected. It is induced experimentally over heating the developing fruits. Foster and Venter, in 1975 found that when plants are supplied with nutrient solution containing potassium, the percentage of green-back can be modified. During very high summer temperature, reducing the temperature can minimize green back.

Collar Rot

This is caused by the evaporation of nutrient solution at the surface of the substrate used for establishing the plant. A temperature of 35°C considered to be the maximum root temperature suitable for most plants growing in hydroponic systems.

Watery Fruit

It is a physiological disorder resulting from an imbalance between plant water absorption and ambient climatic conditions. The massive influx of water into the fruit, due to an excessive root pressure, increases the volume of cells and can sometimes damage them. The organoleptic quality of fruit is then negatively affected and shelf-life is much reduced. Maintaining plant leaf area index at a reasonable level during summer time helps reduce root pressure and minimize the incidence of this physiological disorder. Over irrigation by the end of the day and a strong root system development also favour the appearance of watery tomato due to an elevated root system.

Apical Necrosis

It is caused by irregular watering or excessive salinity

Toxic Effects

Those caused by an excess use of plant growth regulators, may in turn cause malformation in the fruit.

Shedding of Flower or Fruit

This is due to very high temperatures accompanied by low humidity. It may also be the result of low temperature.

Other Fruit Blemishes

The fresh market value of tomato is influenced greatly by appearance, while uniformity of colour is equally important, when fruits are used for processing. Aside from blossom-end rot and sunscald, numerous blemishes and defects of non-parasitic or insect origin are recognized in various parts of the world. Some primarily under glass, others on the outdoor crops.

A disorder reported in temperate climates which has not been named (Hurd and Graves,1983), shows as an oedema in leaves of tomato and other crops grown in NFT and to a lesser extant in other media. Initially, small (2-4mm) circular transparent areas appear in young rapidly expanding leaves owing to the breakdown of the mesophyll cells, The areas coalesce schizogenously within a few days and after a further 10–14 days become necrotic often affecting whole leaves. In extreme cases the symptoms can spread to other parts of the plant and severely reduced yields. It has been noticed (Hurd, personal communication) that the initial symptoms appear 2 –3 days after a period of sunshine, following a spell (5 – 10 days) of dull weather. Such a sequence during the winter and early spring in Northern European Latitudes may cause young leaves to wilt.

Chapter 3
Potato and Peppers

The potato is grown in almost 120 countries of the world and is staple crop in several of them. It is efficient as a producer of high quality dietary nitrogen a very important source of vitamin C but inefficient as a supplier of energy to the consumer.

In addition to viral, fungal and bacterial diseases, potato tuber is also affected by several disorders, which may be physiological or a biotic in nature, due primarily to environmental factors or physical damage adversely affecting plant growth and its market. Hiller *et al.* (1985) classified internal physiological disorders as either major or minor. Major disorders are those occurring most frequently or those which affect a large percentage of tuber production. Major internal disorders include hollow heart, brown center, internal rust spot (also commonly known as internal brown spot and chocolate spot), vascular discolouration and bruising. Minor disorders include translucent end, jelly end, black heart and low temperature necrosis.

Non- parasitic troubles, including simple stem-end browning of the vascular ring; internal brown spot characterized by brown spots scattered throughout the flesh of the tuber; heat and drought necrosis in the nature of a yellowing and browning of the vascular ring and more external tissues; sun burn or greening due to exposure to light; sun scald or the killing of external tissues from exposure to intense heat and light; freezing injury and frost necrosis, evident as ring, blotch or net types of discolouration; hollow heart or central cavity lined by dead-brown cells; and black heart.

Hollow Heart or Hollow Center

It is characterized by an internal split or cavity resulting from rapid growth induced by an abundance of moisture and food material. It is a common and important disorder of potato tubers which may occur in the field as well as during transit and storage. This is characterized by the development of cavities of various sizes in the center of the tubers, usually brought about by irregular growth patterns. The typical form of the complaint is a star- shaped or lens – shaped, cork-lined cavity in the pith, caused by splitting of the central tissue during growth. Although there may be some cell breakage as well as separation of cells when splitting occurs, there is no massive disruption of tissue, the cavity developing as tissue on either side of the splits is pulled apart as growth continues. It is not recognized until the tuber is cut, showing a cavity formed due the death of a small area of pith cells in the center of the tuber, with the result, adjacent flesh cracks as the tuber grows and the hollow area expands. This malady occurs when a period of slow growth, often induced by moisture stress or low temperatures is followed by a period of rapid growth. Cultivars that set large tubers tend to be more susceptive than others. Hollow Heart accompanies excessively rapid tuber enlargement, some cultivars being more prone to the disorder than others. Susceptible cultivars tested were Gladstone and some times Majestic. The typical form of the complaint is a star-shaped or lens-shaped, corked-lined cavity in the pith, caused by splitting of the central tissue during growth. During splitting there may be some cell breakage as well as separation of cells, but there is no massive tissue disruption, the cavity developing a tissue on either side of the splits is pulled apart as growth continues.

Hollow heart may often be preceded by the onset of brown center, *i.e.* necrosis of pith cells resulting in the weakening of tissue and the splitting apart of cells as the tuber grow. Whether preceded by brown center or not, hollow heart is the result of tissue tension associated with rapid tuber enlargement. Nelson and Thoreson (1986) showed that the frequency of hollow heart was considerably higher in large compared with small tubers.

Causes

It is caused by un-favourable oxygen relations. It may occur in the field when the soil temperature rises above 32.2°C during growth and maturity of tubers or it may occur in transit when the temperature inside the carrying vans rises, for some time above 90.2°C. While in storage it occurs when the tubers are stored in poorly ventilated rooms in closely packed conditions. Hollow heart may occur because of breakdown of tissue in the pith at an early stage of growth, possibly as a symptom of potassium deficiency, followed by enlargement of the cavity by pulling apart as growth proceeds. In such cases incipient hollow heart may be detected in cut tubers, at an earlier stage as a small brown area of disrupted tissue. The complaint is aggravated if conditions favouring rapid tuber growth succeed conditions, such as drought, which have caused a cessation of growth. Hooker (1981a) recorded that in some cases up to 40 per cent by weight of the crop could be affected, and the prevalence of hollow heart in some localities has necessitated the development of non destructive methods of detection, such as x-ray examination of tubers immersed in water (Finney and Norris,1973). Certain environmental and cultural practices have also been reported to induce hollow heart, but the evidences are conflicting. However, high soil temperatures and fluctuations in water supply to developing tubers appear to be important and there is some evidence that modifications in nutrient supply also play a role. (Kallio, 1960). High nitrogen applications, particularly around initiation, can induce higher incidences of hollow heart, where as K appears to lower the incidence (Jackson *et al.*, 1984). There is also evidence for an inverse relationship between the severity of hollow heart and soil Ca status (Vander Zaag and Ffrench,1987).

Symptoms

Dark grey to purple or inky black discolouration occurs in the central tissues of the tuber. In advanced stages, the affected tissues may dry out and separate thus forming cavities (hollow heart). The discolouration may extend to the surface of the tuber also. Large tubers are more susceptible to black heart than small ones. In poorly ventilated rooms even low respiration by tubers uses up the available supply of oxygen. This results in discolouration and disintegration of cells due to adverse enzymic action which continues after the supply of oxygen has diminished. High temperature brings about some sub-oxidation by stimulating respiration. The cells of the tuber disintegrate when interior of the potato heap in the store can not ensure a good supply of oxygen.

Control

To avoid losses due to hollow heart, tubers should not be stored or transported at temperature above 32.2°C.

Its occurrence can be minimized by maintaining uniform soil water conditions.

Avoiding over-fertilizing with nitrogen.

Avoid using cultivars prune to hollow heart.

Storage rooms should be well ventilated.

Bags full of potatoes should not be piled upon each other.

Brown Center

It is generally considered to be a precursor to hollow heart although hollow heart does not always arise (Hiller *et al.*,1985). It can be characterized by the appearance of small groups of necrotic cells and eventually by the production of wound cambium around the necroses. Cool temperatures around the time of tuber initiation are reported to induce brown center *e.g.* between 10 –15° C (Hiller *et al.*, 1979) and the severity is increased by high soil moisture.

As with hollow heart, tuber growth rate plays an important role in the initiation of brown center and again the disorder manifest itself in small as well as large tubers. It is also reported that some crops have high incidence of brown center early in the season but show only a low incidence at harvest. This is certainly due to the dissipation of dead cells as tuber grows and not to some reversible phenomenon.

Control

Brown center can be controlled to some extent by applying ethephon (Hiller and Koller,1984).

Growing of resistant cultivars.

Internal Rust Spot

IRS- is characterized by rust coloured lesions, of various size, in the medullar tissues internal to the vascular ring, where as the lesions are generally more pronounced towards the apical end of the tuber.

In addition to calcium, several environmental factors *viz.* restricted water supply and high temperature are responsible for this disorder (Hooker,1981). There is a strong opinion that Ca is the main cause of the disorder. Although many soils have a high Ca content, Ca ions are transported within plants primarily in the xylem stream, a high phosphate concentration precluding substantial transport within the phloem network. Water potential gradients within the potato plant favour xylem transport into the foliage, which has a high Ca content. Tubers are likely to receive most of their water along with sucrose in the phloem mass flow stream. Tubers therefore have a very low Ca content, a content which may be considered marginal for maintaining cellular integrity. Decreasing the supply of Ca to developing tubers increases the incidence of IRS (Collier *et al.*, 1978). As there are incidences where soil applications of Ca reduces the severity of the disorder (Tzeng *et al.*, 1986).

Causes

In hot, dry weather, tubers lose water and probably Ca to the haulms. The absolute concentration of tuber Ca required to induce IRS is uncertain, but it is clearly genotype-dependent (Collier *et al.*, 1980).

Control

Grow less susceptible cultivars

Delayed planting (and other treatments reducing tuber size).

Modify cultural practices to prevent rapid fluctuations in tuber growth rate (Hiller *et al.*, 1985).

Vascular Discolouration

According to Cromack (1981) aberrant synthesis of anthocyanins can occur under certain stress conditions in the flesh of varieties *viz.* Record and Ulster Sceptre, where it is not normally present. Precursors of anthocyanins are present in all tubers but so are inhibitors of synthesis and the purple/red pigmentation occurs when the balance of these components is affected. The pigmentation is largely confined to the center of the tuber and is believed to result from one or two causes. In some it can develop under conditions which check tuber growth, but not in all cases the disorder is associated with greening of the tubers. The disorder seems to be more prevalent when the tubers are formed near the surface of the ridge or when the earthing up has been inadequate. It is reported that irrigations may reduce the number of tubers affected, whereas temperatures less than 10°C, particularly at night, have an adverse effect. High levels of organic or mineral N may also favour pigment synthesis etc.

Control

Provide wider rows

Resort to deep planting or bed system is beneficial.

Black Heart

The potato is affected by a number of diseases marked by a discolouration of the interior tissues of the tuber, but one of these internal necroses which causes a characteristic blackening of the center and sometimes of more external parts has been called the black heart or, less frequently, heart rot. The internal discolourations of the tuber may be caused by either parasitic or non-parasitic influences.

This disorder is caused by sub-oxidation. As whenever potatoes are stored in piles, the air can not get to the center of the pile, with the result temperature at the center rises, which favours the production of this disorder. Even with slight rise of temperature from 3°C to 4° C in not safe for storing the potatoes in piles. In other words it is the warm temperature and excessive soil moisture may give rise to a blackening of tissue in the center of the potato tuber. This disorder affects its appearance, but does not imply pathogenic decay.

In potato different enzymes are inactivated at various temperatures over a range of broadly speaking, 40-60°C, and such temperatures are obviously injurious. The potato, however suffers at temperatures rather lower. If potato tubers are stored in air at a temperature of 35–40°C, they develop, in a few days a condition known as 'Black heart' in which the inner tissues break down, and becomes black. Lack of O_2 is probable initial cause, in that similar symptoms, though more different may be induced by sufficiently lowering the ambient O_2 concentration at temperatures of 5 –20°C. It is also probable that at high storage temperatures accumulation of CO_2 at the center of the tuber might contribute to a damage and to the subsequent decrease in respiration and development of black heart. We could expect over 20 per cent in the intercellular atmosphere during the early stages of storage at 35–40°C, when the respiration was at its peak. It is therefore, caused due to adverse enzymatic reactions resulting from sub-oxidation, poor ventilation and high temperature during storage.

Symptoms

The affected tubers may appear perfectly normal as far as external characters are concerned, but when cut in two they are found to show a browning or blackening of the interior. The discolouration generally starts at the center and progress towards the outside, causing either a star-like radiation or a more uniform advance. If the situation prevails for longer period, the blackening may advance until it reaches the surface. The blackened tissues are in sharp contrast to the normal flesh, similar in consistency to normal tissues or more firm or slightly leathery if they have partially dried. This character should distinguish the lesions from those of either leak or black rot, which are similar in colour but soft and watery. In some cases of black heart of recent origin, the center will be solid; but if of some standing (10 days or more), the shrinkage of the diseased tissues will cause a central hollow surrounded by the black tissue. The extent and the character of the blackened tissue will serve to separate this phase of the disease from hollow heart, in which a growth cavity is surrounded by a narrow zone of brown, oxidized tissue. These blackheart lesions if advance may become wounds, in which rot producing organisms will enter and destroy it completely.

Causes

The disorder is believed to be the result of a derangement of normal physiological processes due to certain environmental factors. It has been shown that blackheart is really due to an asphyxiation of the tissues of the tuber due to lack of sufficient oxygen. This lack of oxygen is much more likely to be effective in causing the trouble at higher temperatures when the life processes are speeded up than at more moderate temperatures, but under these less favourable conditions the disorder may develop if the tubers are not sufficiently aerated. The tubers may suffer from insufficient aeration due to deep piling and under such adverse conditions they may behave essentially like tubers in sealed jars. They sprout feebly or not all, become moist on the surface, discolour externally upon exposure to the air and are often affected with black heart internally. It is the lack of oxygen and not due to accumulation of carbon dioxide given off during respiration.

According to Bartholomew (1915), the heated or asphyxiated tissues of the tuber develop an increased amount of the aromatic amino acid tyrosin. The enzyme tyrosinase is present also, and the interaction of the two results in the formation of a black precipitate, which has been called "melanin" or "humin." The blackening of the tissue is due, then, to (1) the increase in the amount of chromogen tyrosin in free form; (2) the access of an unusual amount of oxygen, due to the killing of the cells; and (3) the accelerated action of the oxidizing enzyme tyrosinase. As a result, the affected tissues undergo a series of colour changes ranging from light pink to coal black. According to Davis (1926) there is an accumulation of CO_2 and a depletion of oxygen in the tissues previous to the appearance of the discolouration. There is a high respiratory activity and exchange of gases fail to keep pace with the respiratory rate. It is also stated that temperatures above 38°C may have a direct effect, since this is the critical point for the maintenance of normal water relations.

Prevention

- ☆ Exposure to high temperatures during storage or transit to the market should be avoided.
- ☆ Over crowding in poorly aerated rooms should be done away with.
- ☆ Potatoes to be stored for more than 6 months or more should not be piled more than 1.80m deep even when the storage temperature is held around 7.21°C; or, in other words, no potatoes should be more than 1.80m from an open-air space.
- ☆ Temperature control alone during the storage period is not sufficient. As potatoes should be stored in such a way that they will be supplied with air, it should be noted that higher the temperature the more the need of a good aeration.

- ☆ Potatoes should not be left long in soil after the vines are dead, especially in regions of high soil temperatures, neither should they be left lying long in the hot sun after digging.
- ☆ Care should also taken not to hold freshly cut tomatoes in large piles, as black heart may develop under such conditions.

Tip Burn of the Potato

This disorder of potato is characterized by burning or browning of the tip and margins of the leaflets under the influence of excessive heat and sunshine. It was Jones (1895) who characterized it as showing the death of the leaves at their tips and margins, which portions dry, blacken and roll up or break off.

Symptoms

The first symptom is a slight wilting and yellowing of the tissues at the extreme tips of potato leaflets or, more rarely, at the margins at some point back of the tip as well as at the tip. This yellowing is followed by a browning and the death of the tissue and the dead area is extended from the tip downward or from the margin inward until in extreme cases entire leaflets are brown and dead. The browning which begins at the tip or margins gradually advances under favourable conditions, downward or inward into the pale-green or chlorotic tissue. This advancement of yellowing may be delayed or checked under unfavourable conditions. The amount of tip burn varies with the position of the leaves, age and maturity of the leaves. As young and upright leaves suffer least, while older and those that occupy a position so that the sun strikes them more nearly at right angles to their surface.

The older leaves droop so that only the tips of the end leaflets are exposed, but those of all of the leaflets may hang down and be struck by the tip burn. The oldest leaves are likely to be under the shade of those from the middle of the stem and so suffer less than those in the middle. Therefore, if they are exposed at all, they succumb readily, and all of the leaflets die in a few days. The earlier attacks are likely to occur on the middle leaves, and after they have become thoroughly scorched and dead the oldest leaves are exposed and in turn succumb (Lutman,1919).

The disorder results in early killing of about 40 per cent of the foliage, thus growing season is cut short and the yield is correspondingly reduced.

Causes

It is attribute to the unfavourable conditions surrounding the plant, especially to the hot, dry weather with insufficient water supply.

Insufficient food supply, insect-pest attack and early blight may aggravate the tip burn. Lutman (1919) was of the view that the principal factors operative in producing tip burn are heat and the intensity of light. The fact that temperature and sunlight are more important than water shortage in producing the trouble can be noted by the behaviour of plants which are not suffering from water shortage. Sun light may act either chemically to cause destruction of important leaf constituents, *e.g.*, chlorophyll, or it may so warm the leaf as greatly to accelerate the water loss (Lutman, 1919).

Control

Since the tip burn is the result if intense light and high temperature in combination with low humidity and frequently of water shortage in the soil. Therefore, disorder may be reduced to some extent provided following precautions are taken.

1. Select late variety if experience shows that early varieties suffer.
2. Light soils should be avoided.
3. Spray to control flea beetles, leafhoppers and other insects.

Black Spot or Bruising of Tubers

Bruising of tubers by roughing handling during harvest or storage results in a disorder known as Black spot. This disorder is characterized by numerous, small, discoloured areas beneath the tuber skin, with few or no outward signs. Tubers that have relatively immature skin, such as some of the early crop potatoes that are often harvested before plants are mature, are most susceptible to black spot. Also an internal browning, black spot develops in vascular tissue 1 to 3 days after the bruising occurs, while as phenolics are suspected as the cause of discolouration.

Control

Growing and storage conditions are to be improved in order to avoid this disorder.

Use resistant cultivars.

Greening of Tubers

This disorder occurs when the tubers are exposed to light, either in the field or in storage, resulting in the formation of chlorophyll in the peridermal layers of tubers exposed to light leads to 'greening' which markedly reduces the product's acceptability.

In the field the major causes of greening are either insufficient cover over the tubers at the planting (Lewis and Rowberry, 1973) or the exposure of tubers following heavy rainfall on light soils. The green flesh that develops is bitter when cooked owing to the parallel synthesis of glycoalkaloids. This comes from a alkaloid compound, Solanine, but may be toxic if eaten. Light intensities as low as 3-11 Wm^{-2} for as short a period as 24 hours could induce greening (Gull and Isenberg, 1958). The development of green colour is influenced by variety, stage of maturity and temperature (Akeley and Houghland,1962). Smooth white skinned cultivars tend to be more susceptible than other types. To prevent this, tubers should be well covered with soil in the field by hilling and must be stored in dark. It was demonstrated that light intensities as low as 3-11 w m^2 for as short a period as 24 hours could induce greening. The development of green colour is influenced by variety, stage of maturity and temperature. Non-greening was found at 5°C and it was extensive at 20° C, the greater effect was observed in immature tubers. The effect of greening was aggravated in potato varieties whose tubers were formed near the surface. Clumping of stems resulted from the use of large seed planted widely spaced. Less severe competition due to wide spacing probably allowed more tubers to set. Some of which forced to the surface, exposing them to light and thus turning them green.

Greening of potatoes is often associated with formation of steroid alkaloids which can cause off flavour on cooking at concentrations of 15-20mg/100g. This concentration is 5-10 times higher than that occurring in normal potatoes. Most of the glycoalkaloids are concentrated in the skin, and in prepared potatoes this is usually too low to cause any nutritional hazards or poisoning. Synthesis of glycoalkaloids is markedly influenced by variety, temperature of storage and intensity of light (Buck and Akeley,1957). Several environmental and management factors elevate levels of glycoalkaloids, including mechanical injury, premature harvest, excessive fertilization and exposure to light (greening).

Control

The plants should be properly hilled to prevent exposure of tubers to light.

Always store tubers in darkness.

Avoid exposure of tubers to direct sun light.

Growth Cracking

Growth cracking is known to be associated with certain cultivars, the reasons for the phenomenon are not well under stood. According to some it is the fluctuating water stress (Iritani,1981) or the conditions which give rise to rapid changes in growth rates (Gray and Hughes,1978). Both these suggestions of the conditions leading to growth cracking may be over-simplifications (Mac Kerron and Jefferies, 1985). Conditions which produce very rapid growth of tubers, which may or may not accompany the relief of water stress can produce the large cleavage cracks observed in cv. Guardian if growth rates are high enough. The relief of water stress can also induce the jagged tearing cracks observed in cv. Record where the preceding stress has been sufficient, along with other conditions, to cause cessation of tuber growth. Fracture cracking of tubers, a form of mechanical damage at harvest, has been considered to be a separate phenomenon from growth cracking.

However, Schoorl and Holt (1983) concluded from compression studies that cracking occurred when the potential energy within a tuber exceeded a critical value. Susceptibility to mechanical damage, including fracture damage, varies with cultivar and is influenced by environment. Lampe (1960) showed that tubers from high rainfall areas were damaged by the lowest rupture force. Similarly, Finney and Findlen (1967) observed that tubers from irrigated plots damaged more easily than those from unirrigated plots, and that high turgor increased damage susceptibility.

Handling during store loading, unloading and grading can add to the damage incurred at harvest. Damage which occurs can be divided into two groups, external and internal. External damage includes scuffing of the skin, cuts or gouges, crushing and splits or cracks. Internal damage is usually of two types, internal shattering or cracking and black–spot. The internal shattering is characterized by cracking of the tuber tissue which may extend to the outer surface. The fractured tissue may dehydrate, with some shrinking and discolouration occurring in the damaged cells around the margins of the shattered tissue. While as the black-spot is an internal discolouration of the tissue which develops some time after the impact. The bruised area is typically blue-grey with diffused edges, although in some tissue it may have well defined margins. The discolouration in both the instances is due to the formation of melanin within the damaged cells.

All types of internal injury result in increased losses during preparation for cooking or processing.

Translucent End Tubers (Sugar End or Glassy End Tubers)

Translucent end appears occasionally in some production areas. In some cases there can be mobilization of starch in the original part of the tuber and translocation of the soluble carbohydrate to the new growth. Here starch is re-synthesized and deposited. The original part of the tuber is at the basal (stolen) end; the new growth at the bud (apical) end. Tubers suffering from this form of second growth are thus depleted of starch at the stolen end, as result of which this end can become more or less translucent. Incomplete translocation of mobilized starch from this region results also in its having a high concentration of sugars. In normal tubers the percentage dry matter at the stolen end is higher than at the bud end, but in translucent tubers the reverse is the case. Iritani and Weller divided tubers exhibiting symptoms of translucent end into apical and basal halves. The greatest difference in dry matter of the halves was 3.2 per cent–16.2 per cent in the basal half and 19.4 per cent in the apical half. If the flesh is noticeably translucent it is a sign that it has been depleted of its starch content as a result of second growth. Depending upon the form of second growth the complaint may occur uniformly through out the tuber or it may be localized at the stolen end. The balance of growth substances in the plant, susceptible to varietal and environmental influence and leading either, on the one hand, to vegetative growth or on the other to tuberization. Second growth is the result of the fluctuations of this balance, from a state favouring tuberization to a state favouring the resumption of extension growth and then back again to a state favouring tuberization.

Causes

It seems to relate to environmental stresses, particularly drought and heat. Affected tubers are often irregular in shape, develop a glassy appearance at the proximal end and may decay in storage. Analysis has shown that there are high sugars and low solids in the translucent area, quite similar to the flesh in the second growth.

Control

Maintain 50 per cent available soil moisture throughout the growing season.

Avoid excessive applications of nitrogen

Jelly End Rot

In this case the extreme stolen end of affected tubers become so depleted of starch that it no longer gives a staining reaction with iodine. The depleted tissue becomes soft and spongy (the tubers then being referred to as exhibiting soft end) and eventually breaks down completely to jelly-like mass which then dries out leaving a shriveled basal tip to the tuber, discoloured by melanin, Although nonpathogenic the root may be invaded by secondary pathogens, the consequent extension of the rot obscuring its original cause.

Chilling Injury

Potato is not normally included in the list of crops susceptible to chilling injury, but low temperature sweetening, though reversible, may be a symptom of a related change, while prolonged storage at 0 –2°C may give symptoms more definitely describable as injury (Richardson and Phillip,1949). Chilling injury may follow prolong storage of tubers at temperature of about 0°C. They include:

discoloured blotches varying from light reddish brown to dark brown, in the flesh of the tuber, diffuse brownish black patches on the skin, and reduced or completely inhibited sprouting. Raison (1974) correlated the injury with a change of phase in the membrane lipids, from the normal liquid crystalline to gel state. The most susceptible cultivars studied by Richardson and Phillip- cv.- Katahdin-showed such symptoms after two months at 0°C and after six month at 2° C.

Freezing Injury

The freezing point–the temperature at which a potato tuber freezes depends on the amount and nature of the solutes dissolved in the cell sap. Freezing points of from–1 to–2.2°C have been recorded. If the tuber has previously sweetened at a low temperature, the freezing point tends to be lower than in the absence of such sweetening (Wright and Diehl,1927).

Tubers which had been only briefly frozen exhibited when cut in half, a blue-black discontinuous ring in the vascular region. More severe injury led to a blue- black necrotic net work in the pith, in addition to necrosis in the vascular tissue. In tubers which had been frozen for more than about one month, the cut surface of the thawed tuber showed diffuse areas of black discolouration. Frozen tubers on thawing present a cheese–like appearance and later the tissue breaks down into slimy watery mass.

Collapse, after thawing of frozen tubers is often, followed by bacterial rotting, which may then spread. High as well as low temperatures can induce reversible changes in sugar content. Thus, exposure for a few days to 30- 33°C may lead to marked increase in the concentration of sucrose.

The level can be reduced again by subsequent storage at 20°C for 10-12 days.

Potato shots are especially frost sensitive and are frequently killed by frost. A frost necrosis of leaves in the nature of minute brown spots on leaves otherwise normal has been attributed to low temperatures (Macmillan,1920). Growers, dealers and consumers are more concerned with effects of low temperatures upon the tubers, either at maturity in the field, during late harvesting or during storage or transit to market.

Three prominent types of injury to potato tubers may result from action of low temperatures:

1. Freezing Solid

Tubers held at a temperature at or below the freezing point for potato tissue freeze solid- either the entire tuber or on one side or end. Tubers close to the surface of ground or slightly protruding are sometimes caught by the sudden early freezes. When frozen tissues are killed and then thaw out, decomposition set in, the tissue becomes softened, the skin is raised and ruptured by gas accumulations and a watery exudates oozes out. The cells beneath the skin are loosened by the solution of the middle lamellae, and a cut surface immediately turns brown. If there is partial freezing, a dark line may separate the frozen tissue from the normal. This is some times followed by decay by either bacteria or fungi (Heald, 2006).

2. Turning Sweet

Tubers stored for a number of weeks at temperatures closely approaching the freezing point, the tissue develops a sweetish taste. This 'turning sweet' has been thought to be due to slight freezing, which is popularly known as "chilling." The sweetish taste may not be good, but it does not cause any lasting injury as tubers when exposed to higher temperatures will become normal again.

3. Internal Frost Necrosis

The tuber if subjected to low temperatures but not sufficiently low or sufficiently long to freeze solid may develop internal discolourations or necrotic areas which are evident on cutting the tuber. Three discolourations have been reported due to low temperatures: (1) the blotch types, they appear as ovoid or irregular patches ranging from a slight metallic tinge to opaque grey, dark brown or almost sooty black and located most frequently in the cortex or in the vascular ring, although sometimes present in the pith; (2) the ring type: this is characterized by lesions in or adjacent to the vascular ring, thus making a continuous or broken ring, narrow and distinct or broader and more diffuse and showing the same shades of colour as in the blotch type; (3) the net type: there is browning or blackening of the fine ramifications of the vascular elements, so arranged as to give a broken, net-like pattern, either exterior to, or within the vascular ring. The internal necroses are most evident towards the stem end of the tuber and in slight injuries may be restricted entirely to that part. They are not indicated by any external markings and the injury is generally evident only when the affected tubers are cut into two. In severe internal necrosis, shriveling may be increased, internal splits or pits may form, while Fusarium dry rot may enter.

High Temperature Injury

Different enzymes are inactivated at various temperatures over a range of, broadly speaking, 40-60°C, and such temperatures are obviously injurious. The potato, however suffer injury at temperatures rather lower. If potato tubers are stored in air at a temperature of 35-40°C they develop in a few days, a condition known as black heart, in which the inner tissues break down and become dark. Lack of O_2 is probable initial cause, in that similar symptoms though more diffuse may be induced by sufficiently lowering the ambient O_2 concentration at temperatures of 5 –20°C (Stewart and Mix, 1917, Kidd and West 1923), and that at high temperatures the symptoms may be prevented or delayed by increasing the ambient O_2 concentration. The sequence of events in a potato tuber at 35–40°C may be possible:

1. A high initial rate of respiration leading to virtually anaerobic conditions at the center of the tuber.
2. Membrane disruption in the anaerobic cells giving the possibility of the sequence of reactions leading to melanin formation, except that no reaction is possible under anaerobic conditions.
3. Progressive slow damage to the respiratory enzymes, leading to a progressive reduction in the rate of respiration and hence to an increase in the O_2 concentration at the center of the tuber.
4. Oxidation of tyrosine by polyphenolase, following the increase in O_2 at the center, leading ultimately to the formation of melanin*.

---

* The reactions which occur in the damaged cells are those involving the oxidation of phenolic compounds such as chlorogenic and caffeic acids and tyrosine by polyphenol oxidase to give coloured end product. A typical example is the formation of melanins from tyrosine by a chain of reactions of which the first two are catalysed by polyphenoloxidase.

It is probable that at high storage temperatures accumulation of CO_2 at the center of the tuber might contribute to damage and to the subsequent decrease in respiration and development of black heart- we could expect over 20 per cent in the intercellular atmosphere during the early stages of storage at 35-40°C, when respiration was at its peak.

Tyrosine

- 3,4- dihydroxyphenylalanine
- 3,4- dioxyphenylalanine
- 5,6- dihydroxydihydroindol-2-carboxylic acid.
- 5,6- dioxydihydroindol- 2-carboxylic acid (red in colour)
- 5,6-dihydroxyindol-2-carboxylic acid.
- 5,6- dihydroxyindol- melanins (black)

Melanin formation is the most important cause of the eventual black or grey discolouration in damaged cells of the potato tuber this being preceded by red discolouration which results from its precursors.

Ring and Net Necrosis

Potato tubers when exposed to freezing temperatures during or after harvest usually develop this malady. In this disorder conductive tissue is affected, either the vascular ring (ring necrosis) or fine vascular elements (net necrosis), followed by discolouration or blackening, thus rendering the tubers unfit for market. The affected tubers show more damage at the proximal than at the apical end. This disorder is avoidable provided the tubers are prevented from freezing.

Ageing of Seed Tubers

The seed tubers of some cultivars such as Ulster Premier stored at too high a temperature, may have a reduced capacity for growth by the time they are planted, giving growers as little potato. Such little 'potatoes' can also be produced instead of normal sprouts on stored tubers. The behaviour is due to the tubers being physiologically old, either because they are actually old or because storage conditions have hastened their life cycle. This condition provides an example of the cyclic pattern of loss and renewal of endodormancy observed by Hogetop (1930).

This system of ageing, associated with the reduced potential for normal growth, had also been observed. Dyson and Digby (1975) were able to arrest the symptoms of aging in several cultivars by application to the sprout tip of 0.01 per cent M Ca SO_4, and concluded that calcium played an important role in the physiology of ageing tubers. There is some suggestion that this could be related in some ways to the auxin/gibberellin interaction to which (Booth, 1963) related diageotropism, but we may also be concerned with maintenance of membrane integrity. The fact that tubers age more rapidly at high temperatures, this could be in terms of day degrees above a specified base temperature. In calcium deficiency dark brown or black necrotic stops may not infrequently be observed just below the tips of sprouts which had previously grown vigorously.

Scorching

If the environmental temperature is high and exposure to light is prolonged, the tuber turns bronze-green and the cells below the bleached area die.

Legginess

The sprouts are long and thin. This is due to excessive heating of the tuber.

Strings of Tubers

The tuber are small and in a row. This is frequent in late varieties due to several successive interruptions to tuber formation.

Cracks and Hollowness

These are due to excessive brusque changes in environmental factors, such as water, temperatures and excess nitrogen availability in later phases of growth.

Excess Lenticel Production

The tuber produces many wart like lenticels due to excess nitrogen supply.

Coloured Patches within the Tubers

They vary in colour and texture and may be blackish and reddish and elongated. They are due to many different causes.

Button Rot of Potato

Shallow surface pitting caused by death and desiccation of areas of surface tissue; frequently followed by fusarium decay.

Peppers

Blossom-end Rot

This malady is characterized by small areas at or near the tip of pepper fruit that become light brown and sunken.

Affected areas on the fruit develops a leathery texture as the fruit matures. Blossom-end rot usually results from an irregular or insufficient supply of moisture and or calcium with the first pepper fruit most often being affected. Reduced soil calcium levels in the soil accompanied with the low soil moisture conditions especially during development of the fruit may restrict the transpiration movement of calcium through plants. Affected areas on one of the lobes on the blossom-end of the fruit first appear water soaked, but soon become dry, light coloured, and papery. Secondary organisms may invade these areas. Cultivars vary in susceptibility to the disorder.

Control

The best control is maintaining an adequate supply of calcium to the plant and avoiding periods of water stress.

Sun Scald

Exposure of pepper fruit to long duration of intense sun light results this disorder. The exposed areas of the fruit become light coloured, slightly wrinkled and papery, followed by the invasion by secondary organisms of the damaged tissues.

Good plant canopy cover will normally provide adequate shade to the fruit and prevent it against sunscald during intense sunlight.

Chapter 4
Physiological Disorders of Cole Crops

They include cabbage, cauliflower, broccoli, Brussels sprouts, khol rabi, kale and collards. All have been derived from wild cliff cabbage called colwarts (*Brassica oleracea sylestris* L.). All these vegetables belong to the family of plants known as crucifers, which also includes Chinese cabbage and water cress etc.

Cauliflower

The most important physiological disorders of cauliflower are whiptail, browning, buttoning and blindness. The first two are deficiency symptoms of specific essential elements, while the last two may result from variety of causes.

Whiptail

Both broccoli and cauliflower have a relatively high requirement for molybdenum. This disorder is due to molybdenum deficiency, which develops in acid soils and the condition can be corrected by the application of lime to the affected soil. Molybdenum is unavailable in vary acid soils and availability is increased by reducing the acidity with lime. This disorder seldom occurs when the soil reaction is pH 5.5 or higher. Some cultivars *viz.* Super snowball type are much sensitive to molybdenum deficiency, than Erfurt types. Usually the leaf blade do not develop properly and may be strap like, and severely savoyed, but in severe cases only midrib develops, which accounts for name whiptail.

Symptoms

In early stages, the deficiency shows up as crinkling, mottling and marginal leaf scorching. In the later stages, plants laminae, resulting in a characteristic 'Whip tail' appearance of the leaves. The growing point is usually severely deformed and does not produce a marketable head and in severe cases there is stimulation of sprouts on the base of the plants. Waring *et al.* (1949) reported that whiptail could be prevented by liming the soil to a pH 6.5 or by applying 1.25kg of sodium or ammonium molybdate per hectare. This compound be first mixed with fertilizer and then applied in irrigation water or applied in water when the plants are transplanted. This malady can be to great extent controlled by spraying the plants in the nursery itself with 1/10 ounce (2.83 g) of sodium molybdate/(0.83 m^2) 2 weeks before transplanting. In order to increase the efficacy of the fertilizers, the best way is to keep the soil to pH 6.5, where it is practical.

Cauliflower responds severely to the deficiency of molybdenum and the damage may be considered. Young cauliflower plants in shortage of this element become chlorotic and may turn white particularly along the margins, they also become cupped and wither. Eventually, the leaf dies and the growing point also collapses. Nitrogen deficiency not only results in buttoning but it also develops the deficiency symptoms of molybdenum. The whip tail develops with high nitrate supply and low molybdenum.

Control

Since cauliflower has high nitrogen requirement, it may be useful to ensure an adequate supply of molybdenum to avoid whiptail. Thus application of 1 kg/ha of molybdenum alongwith super phosphate produces significantly higher yield. Molybdenum deficiency can generally be corrected by liming, foliar fertilizer application or seed treatment.

Buttoning

The term is used to denote the development of small heads while the plants are small. It is considered by many as premature heading. Carew (1947) has shown that curd of normal plants begin to develop as early as that of the button. The main difference is that the developing head of a normal plant is hidden by the foliage until it is of considerable size, while the button is exposed as soon as development begins. The plants that develop buttons are small and have small leaves which do not cover the developing head. However, causes of buttoning can be many *e.g.* over aged seedlings, poor nitrogen supply, wrong cultivars etc. but any check during the vegetative growth of the seedlings may also induce this malady. Carew (1947) has shown that crowding of the plants in the flats and certain other factors that markedly restrict vegetative growth increase the trouble. While, Robbins, Nightingale and Schermerhorn, in 1931 have shown through experiments that nitrogen deficiency is likely to result in buttoning. It therefore, seems probable that nitrogen deficiency is involved in most cases where buttoning is a problem. The cauliflower varieties are very sensitive to temperature and photoperiod. Therefore, it is very important to choose the right variety for the right time. In case of cauliflower there are three main groups and sub-groups within each group. Therefore, early variety if sown late produce buttons head and late varieties if sown early will produce leafy growth. As transformation from vegetative to curding in a variety of cauliflower is dependent on particular temperature, therefore, the check of vegetative growth followed by suitable temperature for transformation to curding may induce this malady. The check in growth may be caused by low nitrogen supply, root injury by insects or by some diseases especially rhizocotonia spp. When early cultivars are planted late and transform easily due to low temperature, over aged seedlings after establishment do not get sufficient time to initiate growth before transformation etc.

It has also been reported that nitrogen deficiency may result in buttoning. Maintenance of rapid vegetative growth through an adequate supply of nutrients and good control of weeds and diseases and pests especially the cabbage maggot, control the disorder to a large extent. Young plants or those that are about six week old when planted in the field, are much less likely to develop buttons than are older plants.

Control

Growers are advised not to plant older seedlings, and to follow practices which will result in rapid vegetative growth e g to maintain an adequate nutrient supply, good control of weeds, insects especially cabbage maggots. It is also necessary to delay planting until weather conditions are favourable for plant growth.

Riceyness

Premature initiation of floral buds on curd giving a velvety appearance is characterized as riceyness in cauliflower, which is considered to be a poor quality from market point of view. According to the observations made at IARI, New Delhi, it has been found that such a disorder may result from any temperature higher or lower than the optimum required for a particular variety. However, there are significant variations among the cultivars in their tolerance to such characters. Such observations have also been noticed by other workers in Europe and America. Varieties differ in the tolerance to this disorder, as some cultivars are more susceptible than others. Wiebe (1975) has also reported that cauliflower cultivars show more braceteate curds at higher temperature than the optimum.

Select a proper variety for a particular season

Use of optimum doses of nitrogenous fertilizers

Plant resistant cultivars to minimize the ill affects of this malady.

Browning

Brown rot or Red rot is caused by boron deficiency. The disorder first appears as water soaked areas in the stem and in the center of the branches of the curd, but the first external appearance is on the surface of the curd. At this stage the trouble is known as brown rot or red rot, as the affected areas change to a rusty brown colour, leading to brown rot or red rot condition. Browning is associated with hollow stems, but hollow stem can occur without browning. The disorder affected curds are bitter in taste when eaten raw or cooked. The other symptoms of boron deficient plants are brittleness, thickening, downward curling of the older leaves and changes in the foliage colour followed by the development of blisters on the upper side of the midrib. The first colour change is a change to a dull green followed by a fading of the green to a greenish yellow colour in a band along the apical margin of the older leaves. While the edges of the older leaves later develop a purple colour, and the greenish yellow band, mentioned above turns orange yellow and extends inwards from edge of the leaf for 1.25–3.75 cm. In severe cases boron deficiency the small leaves and the growing point may die.

Control

The quantity needed to control browning depends on the character of the soil, the soil reaction and the extent of the deficiency. Application of borax at the rate of 10-15 kg per hectare on acid soils will control browning to a great extent. While in neutral and alkaline soils, larger quantities are needed. Because in acid soils boron is applied as borax, a readily available form and there is no danger of any injury to the plants. But much of the boron may be rendered unavailable to the plants if applied in large quantity on neutral or alkaline soils. The experiments by Purvis and Hanna (1940) have shown that the application of common borax (sodium tetraborate, $Na_2 B_4 O_7 10H_2O$) controls browning.

Blindness

Blindness in cauliflower is characterized by plants without the terminal buds and large, dark green, thick and leathery foliage. Axillary buds develop in some cases, but plants do not bear marketable heads. This disorder is very common on over wintered plants and one of the causes is believed to be due to the effect of low temperature on the small growing plants. Blindness occurs in all cauliflower growing regions (Lee and Carolus,1949). It may also occur during early stages of plant growth, if the growing point is damaged by insects, or is damaged while performing cultural operation.

Hollow Stem

In addition to 'whip tail' another important physiological disorder, sometimes found in broccoli and cauliflower, is hollow stem. The disorder is encountered in soils deficient in boron. In heavy soils, particularly with nitrogen, rapidly growing plants may develop hollow stem.

Can be controlled by providing close spacing and proper use of nitrogenous fertilizers.

Symptoms

Curling and rolling of the leaves, deformed foliage, brown curds or brown flower bud and hollow stem centers are the main symptoms of the hollow stem. Boron deficiency is the main culprit of the disorder but all hollow stem is not caused by boron deficiency, as some times a hollow cavity will develop in the stem just below the base of the curd or inflorescence, which is attributed to a very growth rate from excessive nitrogen.

Special Cultural Practices

Certain physiological disorders occur during both the vegetative and reproductive phases of plant development *e.g.* ricy curd, fury curd and leaf curd.

Pinking

Sometimes, exposure of curds to high light intensity, anthocyanin is formed which give pink tinge to the curds. But the occurrence this disorder is not a common phenomenon.

Chlorosis

When cauliflower is grown in highly acidic soils, it develop chlorosis symptoms, there is interveinal and yellow molting of the lower leaves. The affected leaves turn bronze in colour and become stiff. In severely deficient plants abscission of the lower leaves occur and results into small curd formation. Chlorosis can be controlled by applying magnesium oxide @ 300 kg/hectare. Liming of the soil with dolomite limestone to bring the soil pH to 6.5 is an effective control measure. Use fertilizer containing soluble magnesium to it under control.

Frost Injury

The leaves of young cauliflower seedlings turn yellowish-white on both the surfaces. Petioles become flaccid and white. While midrib along with adjacent parenchyma and stem may also be injured. Fully grown curds are more sensitive to frost than smaller ones. However, in cabbage the young leaves are particularly sensitive to frost, as that the center of the head turns brown, while outwardly the head appears healthy. Similar symptoms also appear in Brussels sprouts. Frost injury can be minimized by irrigating the field on anticipating it and by raising the field temperature by creating smoke.

Leafiness

There is formation of small thin leaves from the curds, which reduces its marketability. Usually, extremely small green leaves appear in between the curd segment due to heritable or non-heritable factors. It is being said that prevalence of high temperatures during curding phase aggravates leafiness. Certain varieties are more sensitive to leafiness or bracketing than others. This disorder can be controlled to an great extent by selecting varieties according to their adaptability.

Premature head formation: (*i.e.* before the plant is fully grown). This means the heads are small and malformed. This may be due to excessively low temperatures during the early stages of the plants growth.

Premature flowering: This may be due to high temperatures during flower formation.

Grey dots on the head: Burns caused by the action of the suns rays on drops of dew.

Cabbage

Cabbage like other crops is also affected by some non-pathogenic internal disorders such as Tip burn, black speak and black petiole (mid rib).

Tip Burn

Tip burn is a physiological disease of major importance world wide. Cabbage plants are affected by non-biotic disorders known as internal tip burn, which consists of a break down of the plant tissue near the center of the head. This condition is characterized by tan or light brown tissue, which may later appear dark brown or even black. The affected tissue losses moisture, becomes dry and takes on a papery appearance, which eventually turn brown or black. The affected area may be narrow along the margins of one or two leaves or quite extensive. Tip burn is increased by high nitrogen (Peck, *et al.*, 1983) and high relative humidity (Dickson, 1977).

Causes

The exact nature of the disorder is not known, although calcium nutrition, plant-water relation and growth rate are involved. Lack of adequate calcium, particularly at the margins of inner leaves (Walker, *et al.*, 1961) has been said to be cause in initiating tip burn development. Since calcium mobility in the plants is involved, the affected tissues have very low levels of this element, even in soils well supplied with calcium. Recent evidence suggests that tip burn is the result of a calcium imbalance caused by localized calcium deficiency in one or a group of leaves. Plants undergoing rapid growth, such as exposed to long hours of day light, high temperature or high levels of nitrogen and when transpiration in the plant is restricted, tend to be more susceptible to tip burn development. These plants are unable to deliver calcium to young, actively growing inner head leaves at a critical point in their development (Palzkill and Tibbitts, 1977). An adequate supply of calcium in the soil will not prevent tip burn, nor is foliar spraying with calcium effective because the calcium is fixed by the outer leaves and not translocated to the interior of head. Tip burn is also associated with lack of calcium translocation to the tips of rapidly growing leaves, resulting in death of the cells and consequent browning and blackening of leaf tissue in the head. High nitrogen and rapid growth are associated with the disease. Palskill *et al.*, 1976, reported that factors that produced or increased root pressure deficits are associated with tip-burn. Thus, for effective selection against tip-burn, conditions that stress the plant are required in order to identify resistant plants.

Control

Incidence of tip burn can be reduced by the use of resistant or tolerant cultivars. Fast maturing and high yielding cultivars are generally most susceptible. Maintaining adequate and uniform moisture supply is most effective in preventing its occurrence. Use resistant varieties. Cultivars differ widely in susceptibility, for example, Green Boy and Rio Verde are very susceptible while Titanic, Roundup, and Super-boy are much less so.

Black Petiole

It is also a internal disorder of cabbage. As the heads approach maturity, the dorsal side of the internal leaf petiole or mid ribs turn dark grey or black at or near the point where the petiole attaches to the core. This is a complex physiological disorder in which environment plays an important role in symptom expression.

Black Speck

It is characterized by dark spots that occur on outer leaves or sometimes throughout the head. Its symptoms may not appear at harvest but the initial damage or predisposition likely occurs in the field with the typical symptoms developing during storage at low temperature (Standberg, *et al.,* 1969). Although the actual cause is not known, but it is presumed that high rates of fertilizers, cultural practices promoting vigorous growth and temperature fluctuation are reported to increase the susceptibility (Cox,1977). While high rates of potassium in soil have been shown to reduce the severity of the disorder. varieties tolerant to disorder tend to be promoted.

Bolting

Premature flowering can be caused by, among other factors, exposure of the young plants to low temperatures for a period of time or by periods of drought.

Red Heart Cabbage and Lettuce

Heart leaves of lettuce develop a deep chestnut brown colour and cabbage a typical colour.

Broccoli

Broccoli plants if exposed to periods of stress, ensuing growth in the field after transplanting may result in the formation of small, button-shaped heads and premature flowering (bolting). It is suspected that extremes of temperature, moisture and fertility are the causative factors of these physiological disorders.

1. Internal Tip Burn

Tip burn causes leaf margins to turn brown and leaves to be buried in the head. This has been ascribed to poor water movement with the plant.

2. Broccoli Heat Injury

In many plants of the world, high temperature cause the heads of broccoli to be rough, with uneven head size. Several parameters for screening of heat varieties have been developed.

Brussels Sprouts

Loose Sprouts

This may be due to several causes, such as high temperatures or excessive fertilization with nitrogen.

Chapter 5
Physiological Disorders of Other Vegetable Crops

Lettuce

Lettuce is affected by a number of physiological disorders, *i.e.* disorders in which no organism or virus-like entity is involved, rather these disorders have been ascribed to environmental causes, such as temperature, moisture, storage conditions and other stresses.

Tip Burn

Tip burn of lettuce (*Lactuca sativa* L.) is a problem, causing severe losses in many lettuce growing areas of the world. The unpredictability of tip burn occurrence and the absence of totally effective control measures complicate the problem. Many different causal factors such as soil and environmental parameters have been associated with tip burn development.

It is a physiological disorder of lettuce expressed as a browning of portions of margins of internal leaves of the head, causing severe losses. It is a physiological breakdown of the older internal head leaves and inner wrapper leaves of head lettuce, and occurs wherever lettuce matures during warm weather. It is a necrosis of portions of the inner leaf margins of lettuce. It results in burning or scorching of lateral margins of inner leaves of mature head. Tip burn occurs as the plants approach maturity and is a function of the failure of calcium to reach marginal tissues and is seldom observed in young plants. It may affect leaf lettuce and damage it severely. This physiological break down is accompanied by soft rot, thus renders the lettuce unmarketable, although, it may not be seen externally. Tip burn is characterized by the death and dark- brown discolouration of marginal bands of the large head leaves. While in non-heading varieties, the symptoms appear in the rapidly expanding large leaves as they approach maturity. The mature leaves and the very young leaves are less often injured. Tip burn can prove a serious economic loss to the grower. This disorder is more prevalent in hot weather, and in periods when the plants are growing rapidly. Factors that tend to check the growth rate serve to reduce the amount of injury.

Tip Burn: A Calcium-Related Disorder

It was Kruger (1966) who first prevented tip burn by applying calcium salts directly to sensitive tissue, while Thibodeau and Minotti (1969) not only prevented the symptom by the direct application of calcium salts to leaf tissue, but also induced the injury by the application of ammonium oxalate which, it is presumed, reduced the calcium ion concentration in the tissue by the formation of insoluble calcium oxalate. The dry matter calcium concentrations of 1 per cent or more are found in healthy outer leaves of the lettuce plant, where as the concentration in the inner tip burn susceptible is only 0.1 to 0.2 per cent (Thibodeau and Minotti,1969). They suggested that calcium distribution in the plant and not just availability from the soil dictated whether tip burn would occur. Any factor or combination of factors limiting the supply of calcium to inner leaves in relation to the growth rate of those leaves will induce tip burn. Application of foliar calcium directly to expanding inner leaves is not possible with normally heading crisp head lettuce at late stages of development and this procedure does not work with this type of lettuce (Misaghi *et al.*, 1981b). The development of tip burn results from the contribution of adverse environmental conditions, changes in chemical pathways, changes in growth characteristics and relative susceptibility of a plant to these influences.

Symptoms

The first symptom of tip burn is usually the breakdown and brown discolouration of small spots of tissue near the edge of the leaf. The spots usually occur first between the larger veins, which increase in numbers and coalesce as the disease progresses until the entire marginal band of tissue is killed. The symptoms may appear on only one or two leaves or on most of the leaves in the head and under very unfavourable conditions most of the leaves may be involved. The dead tissues remain dry and are confined to the marginal portions of the leaves unless they are invaded by microorganisms. Various bacteria and fungi may grow in the dead tissues, producing slime. Rotting then starts and the entire head is involved.

Under field conditions, tip burn usually develops somewhat later in plant development when head is well formed and close to maturity. Whenever tip burn develops, it appears as collapsed areas on the young leaves which rapidly becomes necrotic so that further development and enlargement is restricted. Injury begins on leaves that are between one-fourth and one-half mature size and under favourable conditions as small as 1 cm in length. Necrosis begins near the tip on small leaves and on larger leaves, near their periphery.

There is colouration in larger veins at the leaf margins, before the marginal collapse. This darkening is the result from the laticifer enlargement and rupture which releases latex into the surrounding tissue. The released latex causes collapse of parenchyma, occlusion of xylem elements, and most significantly, coagulation of latex within the entire laticifer system between the rupture and the leaf margins (Tibbitts *et al.*, 1965). The marginal tissue, delineated by the disrupted laticifer system, quickly loses turgor and scattered mesophyll cells become necrotic followed by complete collapse and necrosis. When laticifier rupture occurs in small veins, latex may be released onto the leaf surface with a disruption of only adjacent mesophyll cells. There may be no apparent collapse and necrosis of the tissue, but the expansion of the leaf acropetal to these ruptures is less than normal and a slightly malformed leaf results.

The direct reason for tissue collapse in cells with low calcium concentration is not understood, but conclusions drawn are that the loss of membrane integrity is the probable initial stage in injury development. Calcium is known to serve a major role in membrane function (Marinos,1962). It is further reported that excessive membrane leakage, exhibited as water soaking, and rapid movement of potassium ions from cells during early stages of tissue break down are characteristic of apple fruit deficient in calcium (Simon,1978). The fact that 6-benzylamino purine (BA) sprayed onto lettuce plants has alleviated tip burn development (Corgan and Cotter, 1971) provides additional support that alterations in membrane integrity are involved because it is known that BA and other cytokinins have regulatory role in membrane permeability (Ralph *et al.*, 1976).

Tissue collapse in cells with low calcium concentration may be a result of weakness in cell wall structure rather than simply loss of membrane integrity.

Interacting Factors in Tip Burn Development

Many believe that plants making rapid growth are more susceptible than those that have made slow growth. Tip burn appears to be caused by the accumulation of excessive respiratory products in the sensitive tissues during warm nights. Calcium deficiency associated with soil and environmental interaction is one of the causes of tip burn (Burton, 1982).The chelation of calcium by organic acid under high temperature may lead to a local deficiency of calcium (Misaghi and Grogan,1978). Injury seldom occurs at temperatures below 18.3°C or during day light. Environmental factors during growth of the plant are also important, as they favour a rapid, succulent growth which predispose the plants to tip burn injury- excessive soil fertility, excessive soil moisture and warm weather. Environmental conditions that favour rapid respiration and accumulation of respiratory products in the large- head leaves favour the development of tip burn-warm night temperatures and high relative humidity. Anderson in 1946, indicated that water deficiency as shown by high soil moisture tension, is the prime cause of tip burn. He found a highly significant correlation between severity of tip burn and difference between maximum soil and maximum air temperature, tip burn was most severe when the difference was greatest. As greatest difference between soil and air temperature occurs when a cool moist period is followed by a sunny dry period. Similarly, Newhall (1929) and others observed that tip burn was most under such conditions. As during warm dry period transpiration is high and since the soil temperature would rise more slowly than that of the air, the difference between maximum air and soil temperature would be great at the beginning of the warm period. This might result in a water deficit in the plant. Unfavourable seasonal/climatic factors and calcium deficiency are the causes of this disorder. The tip burn development is the manifestation of total calcium deficiency which is increased in plant during exposure to high temperature in growth chamber (Misaghi and Grogan, 1978). This disorder has been found to be associated with the stage of rapid growth (Collier,1983) and conditions favourable for rapid growth increase this malady. Close spacing resulted in lower incidence of tip burn because of slow growth.

A. Insufficient Calcium in Developing Tissues

(*i*) Calcium Uptake from the Soil

Most soils under intensive horticultural management contain large amounts of calcium on the exchange complex that equilibrates rapidly with the soil solution, thereby having the potential to maintain a high concentration of calcium in the solution phase. Its transport to the root surface depends upon the transpiration rate; when the rate of transpiration is high, calcium moves by mass flow, and when it is low, calcium moves by diffusion either in the aqueous phase or by exchange without transfer to the soil solution (Barber and Elgawhary,1974). Calcium uptake is stimulated by nitrate and depressed by ammonium ions (Kirkby,1979), which to some extent explains Wiebe's (1967) finding that more tip burn develops in NH_4^+–N fertilized plants than in those fed with NO_3^-–N. Calcium uptake is confined to the root tip and is transported across the root through the apoplast, which is blocked when the endodermis becomes suberized. Although lettuce develops tip burn injury in well–fertilized soils, it does not exhibit the apical necrosis characteristic of calcium deficiency. Therefore, it is not a short fall in calcium supply in the soils normally used for lettuce production that renders the lettuce crop susceptible to tip burn.

(*ii*) Inadequate Water Movement to Low Transpiring Tissues

Calcium moves only through xylem and is directed through plant mainly in relation to the transpirational water flow (Wiersum,1966), although some calcium moves to the growing point in response to meristematic activity (Marschner and Ossenberg- Neuhas,1977). It moves apoplastically and appears not to be remobilized once it is fixed by the plant. According to Hanger (1979), there is no substantial movement into or transport within the phloem therefore, calcium accumulates in those tissues that lose water most rapidly, *e.g.*, in leaves of lettuce plants that are exposed and transpire freely. The rate of transpiration slows down as the inner leaves become enclosed after the plants have enlarged. As transpiration is negligible when the leaves have become enclosed, because they contain much less calcium than the outer leaves.

The movement of water and calcium into low-transpiring, tip burn susceptible tissue of both cabbage and strawberry plants has been increased substantially in conditions that cause either large diurnal fluctuations in plant water potential (Krug *et al.*, 1972).

(*iii*) Rapid Growth

Frequently tip burn development has been associated with environmental conditions that encourage rapid dry matter accumulation and high demand for calcium in expanding leaf tissue. For example, the symptom develops quickly in young plants when they are exposed to high light intensities and extend photoperiod. It was shown that tip burn develops following exposure to critical amounts of light (irradiance x time) and is independent of photoperiod (Tibbitts and Rao, 1968). Conversely, shading a field grown crop has been found to alleviate tip burn (Collier and Huntington, 1978).

Increasing CO_2 concentrations from 300 to 1500 ppm., air temperatures and relative humidity have been shown to accelerate dry matter production and encourage earlier and more serious tip burn development (Tibbitts and Read,1976, Rao,1966, Read,1972). Plant growth-retarding chemicals applied to foliage are reported to reduce tip burn, because chemical cause decrease in head weight (Borkowski, 1975). While auxin is reported to stimulate cell enlargement and dry matter accumulation is reported to increase tip burn.

(*iv*) Excess Hydrogen Ion Production

Tip burn may also result from excess hydrogen ion production in susceptible tissue, which displaces the calcium from the membrane in susceptible tissue, which displaces the calcium from the membrane phospholipids (Oursel *et al.*, 1973). Excess hydrogen ions have been shown to result from elevated levels of auxin (Rayle and Cleland, 1977).

(*v*) Insufficient Functional Calcium

The application of both oxalate and chelating anions (citrate, fumarate and succinate) to leaf tissue has been shown to induce tip burn-like symptoms. These responses suggest that a reduction in the concentration of calcium ion in tip burn- susceptible tissue either by precipitation or chelation will lead to symptom development.

The other factors that influence tip burn in lettuce are cytokinins, excessive pressure in laticifers and genotypic variations.

B. Cytokinins

Cytokinins, synthesized mainly in roots and transported in the xylem, may be controlling factor in tip burn development because they are known to regulate membrane permeability. It may be a simultaneous increase in both calcium and cytokinin level at the membrane that prevents tip burn development and calcium and cytokinins interact to stimulate ethylene biosynthesis (Lau and Yang,1975). While increase in atmospheric ethylene have been shown to reduce tip burn of harvested lettuce heads.

C. Excessive Pressure in Laticifers

Laticifers which are known to be under high pressure, in young developing leaf tissue of lettuce rupture and release latex in to the surrounding tissue (Tibbitts *et al.*, 1965). Prior to laticifer rupture and tip burn development, laticifers swell, indicating that increases in pressure within these cells may be a determining factor in tip burn injury. (Olson *et al.*, 1967). Turgor pressures within plant cells are a result of accumulation of osmotically active substances. Tibbitts and Read (1967) have shown that the rate of accumulation of metabolites in laticifers is enhanced by increasing radiation levels that encourage more rapid rates of dry matter accumulation and tip burn injury.

D. Genotypic Variation

Varieties differ in their susceptibility to tip burn when grown in same environment, as it has been noticed that fast maturing cultivars are more prone to tip burn, while cultivar Iceberg is resistant and matures early. Early-maturing cultivars have fewer frame leaves, providing a smaller leaf area for transpiration, which may facilitate root pressure flow and movement of calcium to the inner leaves. The extent to which tip burn develops within a cultivar is unpredictable and varies among sites and seasons.

Control

1. Grow resistant varieties like Great lakes, Imperial 456, Progress, Calmar & Salinas resistant varieties often show little or no tip burn, however, under conditions that render susceptible varieties worthless. Crisp head cultivars generally have more tolerance than the other lettuce types.
2. Grow lettuce in cooler parts of the season.
3. Limit the quantity of fertilizer especially of nitrogen
4. Limit the use of irrigation water as the plants approach maturity will aid in controlling tip burn.
5. Calcium sprays may also help reduce the severity of the disorder.

6. This disorder can be decreased by increasing the nitrogen supply (Table 3).
7. This disorder can be prevented or its appearance delayed, by raising the dark period and relative humidity and increasing Ca content in the leaves (Collier and Tibitttts, 1982).

Table 3: Incidence of Dry Tip Burn Harvested at Three Planting in 1990 as Affected by Cultivar and Total Nitrogen Supply

	Tip Burn Incidence (%)		
	Early	*Midseason*	*Late*
Cultivar			
Marius	32a	15a	48a
Saladin	24a	5b	63b
Nitrogen	(Kg/ha)		
50	30a	12a	72a
100	29a	12a	60ab
150	30a	7a	50bc
200	22a	8a	41c

Mean separation among cultivars, nitrogen supply and plant age within columns by LSD at p=0.05

Source: Ref. Sorensen *et al.*, 1944.

Red Heart of Lettuce

This physiological disorder is characterized by the chestnut brown discolouration and breakdown of small, inner head leaves. The outer leaves may appear normal or they may develop numerous, elongate, brown pits on the midribs and veins and sometimes on the tissue between the veins. It often occurs in transcontinental shipments of lettuce, especially in spring.

Causes

The cause appears to be lack of sufficient oxygen, which results from poor aeration or prolonged exposure to low temperatures during shipment and storage. It also results from bacterial rot of the outer leaves, which occurs in shipments without adequate refrigeration.

Control

It can be prevented by providing adequate aeration and prompt and continuous cooling of lettuce to 3.8°C to 4.9°C. in shipping containers, refrigerator cars and terminal storage and by prompt movement of the produce from the grower to the consumer.

Premature Yellowing

Premature yellowing, rib blight and several other physiological disorders occur in head lettuce in the field or during shipping and marketing. The disorders occur more commonly during spring and is caused by various physiological disturbances.

Symptoms

Premature yellowing is associated with poor development of the root system and the production of small infirm heads resulting in the low yields of poor quality. Yellowing has been associated with poor aeration, excessive soil moisture and the accumulation of salts in the root zone. These adverse conditions are due to untimely application of irrigation water and the compactness of the soil formed by the farm machinery.

Control

The losses can be reduced by avoiding use of heavy machinery, especially in wet soils.

Avoid application of excessive irrigation water.

Big Vein

This disorder is characterized by vein clearing and apparent stiffening of leaves. The latter gives the plant a bushy appearance. These symptoms are un-slightly and may also be associated with delay or prevention of heading. It is caused by virus like agent introduced in to the plant by root feeding fungus (Olpidium brassicae (Wor) Dang.

Bolting

This disorder is observed in plants exposed to warm weather usually above 29.4°C as they tend to flower prematurely, thus making the head unmarketable. Sometimes the head may appear normal, but it may have the elongated core and stem inside. Long days, hot and dry weather conditions are reported to induce premature flowering and seed stalk formation. Varieties like Empire, Ithaca and Great lakes are more resistant.

Russet Spotting

Although it may occasionally appear in the field, but russet spotting is primarily a post harvest disorder. This disorder is characterized by the occurrence of reddish, tan, olive, and/or brown elongated pit like spots on the mid rib of leaves. Russet spotting is expressed as small, sunken, rust-brown spots, which appear on ribs of outer head leaves and may progress to inner leaves. When severe, the spots may coalesce. It may occur on both sides of the midribs, and in severe cases the spot may become quite large and appear on the leaf blades as well. It is a post- harvest disorder characterized by localized, spot-like lesions that may start either in the epidermis or in the mesophyll which, in advance stage, may show discolouration of vascular tissue and collapse of mesophyll cells resulting in pit-like depression. Russet spotting is induced by ethylene produced either by the lettuce itself of by other ripening produce, or from outside sources. Ethylene is supposed to be main cause of the disorder, as concentration as low as 0.5ppm may induce spotting in lettuce. Lettuce if held between 3 and 10°C or during prolong storage may also induce russet spotting. According to Rood (1956) the incidence is increased on more mature hard heads and on heads kept at higher- than optimum temperature (0° C). This disorder can be prevented by storing at 0 – 2.5°C and away from ethylene. Russet spotting occurs when lettuce is exposed to ethylene gas. Ethylene is produced by fossil fuel driven machinery in cooling and storage areas, and by several kinds of ripening fruit such as tomato or melon either in the transport load, on the market shelf or in the refrigerator.

Beraha and Kwoleh (1975) found that the incidence of russet spotting was higher on desert–grown lettuce than on coastal lettuce, possibly because the desert lettuce was firmer. The difference also may have been due to the different cultivars grown in the two areas. Lipton, in 1963 found that high temperatures 9 to 14 days before harvest lead to higher incidence of this disorder. In recent study, Morris *et al.* (1978) showed that russet spotting can be induced by ethylene at 0.1 ppm in the atmosphere. Development was accelerated at temperatures over 5° C. The incidence is higher on susceptible cultivars, over mature heads and when the interval between harvest and consumption becomes excessive. Heads in good condition produce little ethylene. Heads damaged physically or by pathogen may produce ethylene at a high rate.

They further studied the environments in which ethylene may be produced at high levels. These include (1) cold storage in which there was forklift trucks with full emission, particularly from propane, (2) retail storage rooms with ripening fruits, (3) the home refrigerators, also with ripening fruits. Damage due to ethylene can be avoided by using non-emitting fuels in forklift trucks, air flushing of confined spaces, wrapping of lettuce in polythene or by the use of gas–capturing devices.

Other factors influencing the development of the disorder include temperature, O_2 and CO_2 levels, maturity and cultivar. According to Ke and Saltveit (1986) calcium and 2, 4-D inhibit russet spotting.

Rib Discolouration

Some times called rib blight or brown rib, occurs on the midrib of the outer head leaves, usually where the rib curves. It appears on the inner (Adaxial) surface, as yellow, brown or black streaks along the midrib and secondary ribs of cap leaves and of leaves just below the cap leaf. These occur on mature heads of crisp head lettuce just before harvest; the disorder has also been anecdotally reported in other types of lettuce. The discolouration is at first yellow or tan, becoming black. The cause is un-known, but the discolouration appears to be favoured by higher temperature (Lipton, *et al.*, 1972). It is characterized by yellow to black lesions that discolour the cap leaf and the next few inner leaves. Disorder occurs when lettuce matures during hot weather. Refrigeration will reduce the decay development in the injured tissue to an great extent. The symptoms of the disorder do not materially change after harvest, but may serve as sites for secondary infections.

However, there is no known cause, but may be associated with high day and night temperatures.

Pink rib is a field disorder of crisp head lettuce (Marlatta and Stewart 1956), and may occur on any other type of lettuce. Pink discolouration may appear on the main rib as the lettuce reaches maturity, although it is more commonly found in over mature lettuce. No organism has been known associated with the disorder. However, pink rib disorder may be accelerated in development at higher than normal storage or transit temperatures, and varieties differ in the development of the disorder. This is therefore characterized by discolouration of the large mid ribs to light or dark pink. In severe cases, small veins are also discoloured.

Internal Rib Necrosis

Considered a post harvest disorder, but it actually occurs in the field. It is also a seasonal, cultivar specific problem, occurring on mosaic infected 'Climax' in the mid winter desert season. It is a grey or black discolouration of lower mid rib near the base of the leaf (Johnson *et al.*, 1970; Coakley *et al.*, 1973).

But rusty brown discolouration and internal rib necrosis are associated with lettuce mosaic, and both are essentially controlled when lettuce mosaic is controlled through use of mosaic free seed in desert plantings of 'Climax.' Further guarantee of control will occur when Climax is replaced by another cultivar resistant to mosaic or at least to the subsequent effects.

Pink Rib

It is a field disorder of crisp head lettuce (Marlatt and Stewart, 1956).This disorder is characterized by a discolouration of large midribs to a light or dark pink. In mild cases of pink rib, the mild rib of the outer leaves of the head shows a diffuse pink discolouration near the base. This disorder may be quite severe, affecting all but the inner leaves of the head and extending well up the mid rib into the larger veins as well. Pink discolouration may appear on the main rib as the lettuce reaches maturity, but it is more commonly found in over mature lettuce. No organism has been recorded associated with this disorder. However, it may be accelerated in development at higher than normal storage or transit temperatures.

Rusty Brown Discolouration

This is a post-harvest physiological disorder, characteristics reddish brown discolouration of the midrib and near by tissue on the outer leaves. Ceponis *et al.*, in 1970, studied and named the disorder and stated that it occurs on 'Climax' a mid winter cultivar grown in the imperial valley. They further reported that 90 per cent of the lettuce arriving in the market of New York during February was affected with this disorder. While Coakley *et al.*, in 1973 induced its symptoms in the heads of lettuce, inoculated with lettuce mosaic virus at relatively late stages of growth followed by storage at 1°C.

Brown Stain

This disorder was first observed in New York market during 1965. Disorder develops when lettuce is shipped in excess CO_2 atmosphere, as small sunken lesions with dark edges occur on either leaf surface, usually near the leaf base and on or near the midrib. These lesions are water soaked when young, but become darker and may coalesce when the injury is severe. Heart leaves in injured head may have reddish brown margins or the entire leaves may be discoloured (Stewart *et al.*, 1965; Lipton *et al.*, 1972).

Brown stain is a disorder caused by excess CO_2. Gas is a normal product of respiration, but may reach higher than normal levels when there is little or no gas exchange, as in storage rooms. The effect is expressed in two ways. While as brown lesions appear about the midrib of the enclosed leaves, but not leaves closest to the interior of the head. These are usually about 5mm x 12mm, but may be smaller or may coalesce to form larger lesions and if the lesions are small they may be confused with russet spotting. Excess CO_2 may cause a reddening of the smallest interior leaves. Brown–stain expression may be increased at a lowered O_2 level or by increased carbon monoxide (CO) as well as high CO_2 (Kader *et al.*, 1973). This may occur when lettuce is kept under modified atmospheres to reduce discolouration.

Brecht *et al.*, 1973 found that the incidence of brown stain increased as the level of CO_2 increased from 1 to 50 per cent and as O_2 decreased from 21 to 1 per cent. Although CO alone has no effect. Kader *et al.*, 1973a found that where CO was present, brown stain increased as CO_2 level increased, regard less of O_2 level.

Brown stain may be controlled by maintaining an adequate O_2 level, a low CO_2 level and/or low CO.

Low Oxygen Injury

This is a post harvest disorder. It was rated by Beraha and Kwoleh, in 1975. It occurs on lettuce shipped in a low oxygen atmosphere.

A number of atmospheric gases like ozone, sulphur dioxide, nitrogen dioxide and peroxyacetyl nitrate (PAN) have shown to have toxic effects on lettuce. Plant damages due these toxins can include visible discolouration, pitting or necrosis of leaves, which on lettuce can render the entire crop un-saleable and even at relatively lower levels can affect growth and yield.

Bolting

This is due to high temperatures

Poor Head Formation

This is due to deficient fertilizer or excessive fertilization with macronutrients.

Other Disorders

Carbon dioxide level if exceeds 1-2 per cent, a brown strain may develop within a week at a temperatures usually encountered during marketing of the crisp head lettuce. This type of injury is certain if CO_2 level is more than 4 per cent. Where as O_2 below 0.5 per cent may induce a reddish–brown discolouration of the heart leaves. Some non-parasitic troubles like crushing and bruising, russet spotting and brown stain may develop due to excessive CO_2 and pink and rib discolouration may also occur.

Wrapper and cap leaves on the head develop patches of shining or water soaked dead grey patches. Young leaves may turn reddish brown (Lipton *et al.*, 1972).

Several problems often originate after harvest, during handling, storage, transportation or marketing. The most important of these are russet spotting, brown stain, bacterial soft rot and grey mould.

Carrot

Forked carrot roots may be due to rocky or stony or heavy soil. Carrot plants with full foliage top grown and small or limited roots can result from seeding too close in the field and not thinning. Excessive nitrogen fertilization can also contribute to extensive top growth at the expense of root growth. Poor-tasting carrots (tasteless, woody or bitter) often result when carrots are grown under inappropriate environmental conditions while the root is maturing. Pale yellow colouring of the root also can be an indication of poor environmental conditions during growth. Splitting and cracking of carrot root is a major problem in many carrot growing areas. Although cracking is controlled by genetic factors, a number of other factors such as high nitrogen and chloride levels in the soil may be involved (Bose and Som, 1986).

Cavity Spot

Calcium deficiency in carrots causes a disorder known as cavity spot (Maynard *et al.*, 1963), in which a cavity appears in the cortex and in most cases the subtending epidermis collapses to form a pitted lesion. It may be induced by excess potassium uptake during the ontogeny of the plant. Increasing calcium accumulation reduced the incidence of cavity spot in carrots. This disorder can be brought under control by application of calcium. Increased calcium level in the growing medium results in increase calcium accumulation in the plant and a significant reduction in the incidence of cavity spot (Maynard, *et al.*, 1961). Apply balanced quantity of nitrogen and do not allow carrots to grow for longer periods of time without water.

Scab Spot Complex

This disorder is related to climatic, nutritional and genetic factors, causing considerable losses in carrots. This disorder was originally attributed to bacterium and was called bacterial blight, but is now known as a physiological disorder, although a few species of bacteria may be found in the lesions. This disorder causes considerable market losses in carrot growing areas of South western U.S.A.

Splitting

Splitting or cracking of carrots is a major problem in many carrot growing areas. A common disorder noticed in carrots, in which roots crack. A number of factors, like heavy side dressing of nitrogen fertilizer at early stages of growth and boron deficiency are responsible for triggering this disorder. According to Boss and Som,1986 high nitrogen and chloride levels in the soils may be involved. However, genetic factor (Bienz,1968) can not be ruled out. Therefore, low application of nitrogen and use of resistant varieties are the alternative to over come this disorder. This is a major problem in many carrot growing areas. Wallace *et al.*, 1942 showed that splitting tended to reduce by low nitrogen and to increase by chlorosis. It may also be caused by a fluctuating water supply, especially when there is a heavy rainfall after a long spell of drought, the inner flesh of the carrot expands faster than the toughened skin, causing the skin to fissures, sometimes roots split often exposing core.

Always apply a balanced quantity of nitrogen and do not allow carrots to grow for longer periods without water.

Bitterness

It is a storage disorder, where ethylene produced causes an increase in the total phenol content in carrot roots, resulting in the formation of new compounds like isocoumarin and engenin, which are responsible for the formation of bitter flavour.

Drought

The roots are tough and stringy

Cracking of the Roots

This is caused by sharp changes in soil moisture.

Cleft Roots

This is the result of inadequate soil preparation, especially in stony soils.

Bolting

This is the premature flower production.

Beet and Chard

Beet is grown all over the world both as salad as well as a canning crop. Like other vegetables it is also subject to some disorders and the most important disorder is Internal black spot. It is a major disorder in many beet growing countries of the world.

Internal Black Spot or Brown Heart

Symptoms

Beets require higher amounts of micronutrients, especially boron. Therefore, any deficiency of this nutrient may result this physiological disorder. Boron deficiency may cause a physiology disorder in garden beet, which is known as Internal brown spot or brown heart or heart rot (Walker, 1939). Internal black spot is a boron deficiency syndrome of garden beet and chard which is characterized by irregular, black spots throughout the interior of the root. Hard or corky black spots develops on the roots, especially in light- coloured zones of the young cells and tissues. Black spots may also form on the surface of the roots, usually at the point of greatest circumference. The plants usually remain dwarf or stunted. The leaves are smaller than normal. The young unfolding leaves fail to develop normally and eventually turn brown or black or die. The leaves may assume a variegated appearance due to development of mixture of yellow and purplish blotches over parts, while the stalk of such leaves shows longitudinal splitting. The young leaves are first to show the symptoms as they become redder and narrower than the normal and are stunted. Some times the leaf blades roll down and die early. Growth of the adventitious buds follows a similar pattern, and a cluster of dead leaves is left at the crown. The most conspicuous symptom on garden beet is the spotting of the root. The spots are black, corky areas, masses of dead cells that might be very small or involve much of the root. The spots inside can not be detected unless the roots are cut.

The first effect of boron deficiency in beet is one of increased cell division and growth, followed by death of the tissue and a reduction in the conductive tissue of the plant resulting in the death of the entire plant. The most certain sign of the disease is the necrotic cross hatching of the tissue on the concave surface of the leaf petiole. As the young unfolding leaves fail to develop normally, turn brown or black and die. If boron deficiency remains unabated, the dormant buds in the axils of older leaves may grow out and die. As this process continues, the heart of the beet top becomes a rosette of small, dead and desiccated leaves, this phase being known is heart rot. Within the fleshy root of beet break down occurs in one or more of the cambium rings. When the brown necroses occurs near the surface, the break down may lead to a canker and secondary organism may enter. While in table beet the necrotic areas are nearly black and are very conspicuous in the canned product. The root phase in sugar beet is referred to as dry rot and that in table beet as internal black spot.

The effects of boron deficiency are evident before macroscopic signs appear, as the meristematic tissues are the first to show the micro symptoms followed by the discolouration of the cell wall and the granulation of protoplast. Where as in cambium, cell division continues with little differentiation in to xylem and phloem elements, and many cells become abnormally large, which leads to ineffective translocation and eventually distorted growth. In regions of abnormal cell activity death of cells comes next in necrotic group which eventually reach a size sufficient to be macroscopic. The cork- cambium zones commonly build up in healthy cells surrounding the necrotic area. In beet root where many tertiary cambium rings develop in the peridium the black spots are found in such rings.

Causal Factors

Boron deficiency may cause the physiological disorder in garden beet which is known as Internal black spot or brown rot or heart rot (Walker, 1939). Many soils are naturally deficient in boron where in others is in the form of unavailable to the plants. While as a small amount of the element is needed by the beet plant, less than the minimal requirement leads to the break down. Boron is deficient in alkaline soils or in soils high in calcium, some amount of it is also fixed in these soils. Some times over liming may also bring on the disorder.

Control

The disorder can be controlled by applying 25 kg of borax per hectare to the soil along with the fertilizer. But when such fertilizer is applied along with the seed, it is necessary to drop the fertilizer mixture in a separate furrow about 5 cm away from the seed furrow. Broadcasting of the borax is the safest and also effective. Resistant variety like Long Dark Blood should be preferred. Detroit Dark Red variety of beet shows comparatively less symptoms of boron deficiency. If seeds are sown early and temperature is high, beet roots produced are often coarse and with woody flesh and dull in colour. If too much fresh manure or compost is added shortly before planting or growing in rocky or heavy soils, poorly formed hairy roots develop.

Tip Burn

The leaves are distorted in shape and necrosis appears along the margins of the blade, often extending to the tip. The leaf blades may cup downward or upward as the necrotic margins where growth has practically ceased become stretched. In extreme cases truncated blades with blackened edges or bladeless petioles with blackened tips occur, chlorosis and nectrotic spots in the blades sometimes develop. On seed plants, the symptoms just noted are supplemented by blackening of the tips of the apical flowering shoot. It is common for plants in which tip burn develops to recover completely (Fife and Carsner,1945).

Causes

It occurs after periods of low light intensity especially in foggy weather. Plants which have had relatively high nitrogen fertilization, appear to be more susceptible to this injury. Strains and varieties of sugar beet show variations in hereditary tendency to develop the disorder, when the optimum conditions prevail. It is envisaged that accumulation of nitrogenous constituents in roots in excessive concentrations brings about this disorder. It is reported that certain factors associated with photosynthesis appear, to neutralize the toxic effect. In support of this theory is the fact that, when older leaves of the plant were exposed to intense sunlight while young leaves were shaded, the disorder did not appear in the latter.

Brown Heart

A disorder in which young unfolding leaves fail to develop normally and the plants eventually turn brown or black with rough, unhealthy and greyish coloured roots. The disorder is attributed to boron deficiency and can be overcome by applying boron @ 11.2 kg per hectare to the beet field.

Bean and Lima Bean

Baldhead and Sneak Head

Baldhead is the result of mechanical injury to bean and lima bean. Seed during threshing, seed with severely mutilated coats are removed in the milling and cleaning processes, while those which have been injured only internally are ones which produce baldhead and sneak head plants. If hypocotyl is cracked below the plumule, the later drops off, and the emerging plant fails to grow. Normally because of the lack of the normal growing tip. Such plants are designated as baldhead plants. The plants may die or belated development of buds at the cotyledonary nodes may result in no growth, in distorted growth, or in a abnormal system of adventitious roots.

Snakehead, a malady brought about by chewing of germinating seedlings by seed corn maggot, has symptoms which develop those of baldhead. In general the damage from baldhead is greater in seeds low in moisture. The injury increases with increase in speed of the cylinder in the threshing machine. Subsequent handling and cleaning process may also be source of damage. Varieties differ in amount of damage, some requiring more mechanical agitation to free seeds from the pods. The chief means of control is in attention to speed of the cylinder at threshing. Threshing have been improved in various ways to effect the damage, such as providing rubber rollers to reduce stress on seed.

French Bean

Delayed Flowering and Pod Development

French bean being a cool season crop is sensitive to high temperature, as flower initiation and pod development is greatly affected under subtropical temperatures especially when the temperature is below 10° C or is above 21°C. The pods even if are produced are small and misshaped. At temperature above 35°C and above there is blossom drop and the ovule abortion.

Cotyledon Cracking

Transverse cotyledon cracking takes place when dry seeds of beans are sown in wet soils. Therefore, resistant varieties with hard seed coat and optimum seed content, and planting crop at suitable time are essential to avoid this disorder.

Sun Scald of Beans

The first indications of the disorder are very tiny brown or reddish spots upon the upper or outer valve away from the center of the plant. These spots gradually lengthen until they appear as short streaks running backwards and downward from the vertical toward the dorsal suture. In 2 days, the spots have increased to areas of brown water-soaked tissue some times slightly sunken. If the spread has been rapid, the colour is a gold brown, sometimes tinged with red, extending over a majority of the exposed surface and sometimes over all of it. On some varieties, the entire exposed surface does not become covered, but spots 3 to 4 millimeters in diameter grow to be the largest, while new spots are constantly appearing. Often small spots coalesce into larger ones, giving them an irregular shape. Eventually this spotting may appear on the underside of the pod but always in lesser quantity (Macmillan,1918).

The sun scald of beans cause little if any loss in a seed crop; but in severe cases, white-skinned varieties may have the seed coats slightly stained, thus affecting quality. There is no loss of vigour or reduction in yield. The disfiguring lesions on pods of string beans are of importance as affecting quality, and it seems probable that under certain conditions they may be the avenues of entrance of certain parasitic organisms.

Yam Bean

Cracking of tubers is the main problem encountered by the growers. Availability of proper soil during the growing period (45-90days) checks cracking. Application of potassium reduces cracking of tubers. Any delay in harvesting also causes cracking in tubers.

Heat Injury and Sun Scald

When beans are grown in regions where young plants are exposed to high mid day temperatures, heat injury lesions, consisting of a constriction of the stem, may appear near the soil line. It is particularly noticeable in light sandy soils. Sunscald is the name applied to a disorder which affects all above ground parts of the older plants. It is attributed to the effect of intense sunlight rather than to heat injury. It appears on parts of the plant exposed to the sunrays. Small brown patches appear between the veins of the leaves, often extending to large patches. Defoliation may follow. Spots appear on the exposed sides of the pods, first water soaked, then sunken and tinged with red. As the spots enlarge, they may develop into short streaks. The symptoms may be confused with those of bacterial blight.

Runner Beans

Pod Ageing or Discolouring

This occurs in the cold storage of produce from old plants.

Stringy Pods

This is due to low temperature.

Shedding of Flowers and Young Pods

This is caused by very high temperatures and very low humidity.

Celery

Cracked Stem

Cracked stem is some times called brown check, is related to a deficiency of *boron*. There is brownish mottling along the leaf margins, followed by the brittleness of the petiole. The affected tissues collapse, become light yellow and later form a corky layer. The disorder begins as a crack of the inner petiole surface below the first leaflet and extends partway down the stalk. The exposed tissue becomes brown and susceptible to disease infection. Crosswise cracks appear in the outer layers of the petiole and tissues surrounding the cracks turn brown. The roots also turn brown and die. Disorder is caused by *boron* deficiency, which is accentuated by high rates of *nitrogen* and *potash*. The disorder can be controlled by soil application of 1.25 to 2.5kg of boron per hectare before planting or spraying of 625g of boron per 1136.5lit of water per hectare or application of commercial borax at 12kg per hectare in the soil at the base of the plant. It may be applied in dry form or can be mixed with the fertilizer. More amounts of borax would be required on neutral or alkaline soil than on acid soils

Resistant variety like Tall Utah 52-70 should be planted if there is boron deficiency. The reaction of cultivars also is controlled genetically, with inefficient cultivars recessive to efficient. It can be controlled by application of boron. Boron deficiency results in cracked stem with lesions on the inner and outer surface of the petioles or over the vascular bundles. The adjoining epidermis curls on words followed by a dark brown colour of the exposed tissue.

Black Heart

It occurs in most regions where celery is grown. It is the young leaves that show tip burn symptoms first, then spreads quickly to most of the heart tissues. This is followed by drying, blackening and in severe cases the killing of the entire heart.

This is caused by the calcium deficiency in the young leaves, which causes the black heart *i.e.* inner leaves and the growing point to turn black and the affected tissues are invaded by the organisms *viz.* bacterial soft rot, resulting in black, water soaked centers. The malady is not confined to the inner leaves but may affect the outer leaves also which turn yellow. The disorder is very similar to tip burn in lettuce and cabbage, appearing when weather conditions favour a sudden growth surge and available calcium can not meet plant requirements. Disorder is reported to be accentuated by high temperature, rapid growth and water stress. It is a growth related disorder, as large plants with high nitrogen rates are more susceptible. The disorder is more serious in regions where water contains high levels of sodium. This disorder can be controlled by spraying 6.5 to 12.5kg of calcium chloride in 1136.5 lit of water per hectare or calcium nitrate 6.5 to 18.75kg per 1136.5 lit of water per hectare to the growing point. The disorder may appear in the field as well as in storage. Black heart or tip burn is a physiological disorder and appears as blackening and dying of tissue in the heart of the plant. It is severe in hot weather accompanied by high soil moisture and high humidity.

The chemical analysis of affected tissues of the young leaves showed that the calcium content was markedly lower than that of similar tissue of plants unaffected by the disorder. Therefore, it seems that black heart may be caused by a lack of balance or antagonism between cations (sodium, potassium and magnesium) and calcium.

Control

- Improved fertilization and regular irrigation reduce the incidence.
- Spraying of calcium nitrate or calcium chloride 0.10 molar solution on foliage is also effective.
- Adequate irrigation during the growth can prevent black heart.

Pithiness

Pithiness is the break down of the thin walled parenchyma cells that form the major part of the leaf stalk, resulting in softening of the tissues and the hollow cavity. It first develops in the out petioles, which has been attributed to the rapid growth, particularly due to high nitrogen rates. Some cultivars are more susceptible to than others. Emsweller (1932) distinguished between two types of pithiness, one clearly hereditary, the other presumably affected more definitely by environmental conditions. The hereditary type affects all petioles of a plant and may be eliminated by careful breeding. Strains very in susceptibility to the second type, which is to be found in the outer leaves and develops as the plants chiefly mature.

Chlorosis

This is characterized by yellowing of older leaves due

to magnesium deficiency. It occurs in cultivars that are inefficient with respect to magnesium uptake and metabolism. It is due to a single recessive gene, but high calcium in the plant will intensify the disorder. This disorder can be ratified by spraying magnesium sulphate (Mg SO_4) at 25kg per hectare on the foliage every two weeks, while soil application is not effective. Cultivars like 'Summer Pascal' are fairly resistant to this disorder.

Pencil Strip

This disorder is associated with high soil phosphorus, which shows narrow brown lines on petioles and its harmful affects can be minimized by judicious application of phosphorus fertilizers.

Hollow Leaf Stalks

This may be due to a period of frost or to over maturity.

Bolting

This is premature flowering, which may be caused by a period of two weeks with temperatures below 10° C when the plant is young.

Hypocotyl Necrosis

This was found to be associated with low calcium content in seeds. Soils rich in calcium and magnesium can offset the problem.

Onion

Bolting

Onions like other vegetable crops also suffer from certain disorders. Bolting is one such, which occurs under certain conditions and produce premature seed stalks, with the result such bulbs become light and fibrous and have very poor keeping qualities. Bolting or sudden breaking of the normal life cycle may be due to certain factors such as generic factors, changes in temperature, poor seed quality, poor soil and cultural practices affecting the growth. The other factors which are responsible for bolting are relative length of day and night, spacing and seedling size (Pandey *et al.*, 1990). Singh and Dinker (1989) were of the view that very high nitrogen doses reduce bolting, they further reported that bolting increased with an increase in bulb size. Bolting of seed stalk formation during the first year of growth of a biennial plant can be a problem during March-April (Spring). The size of the over wintering plant and the exposure to low temperatures appear to be most critical factor determining whether the plant will bolt. Sometimes extended warm period following planting produces a large over wintering plant (over 0.635 cm shank diameter), which results in a high percentage of bolting when exposed to extended temperatures below 10°C. Bolting may be provoked by several causes, such as sowing too early, climatic and cultivation factors, *e.g.*, excess nitrogen and excessive irrigation etc.

Splitting and Doubling of Bulbs

This malady is reported if onions are grown under adverse conditions like hard soils or in fields with poor fertility. Mechanical injury during cultural operations may also cause splitting and doubling in bulbs. Choudhary (1967) reported that water scarcity at the initial growth stage and irrigating after a long spell of drought may also lead to splitting and doubling of onions. The other causes could be erratic moisture and\ or fertilization during growth. If onion bulb growth is slowed or checked from insufficient soil moisture during the growing period, the outer scales will begin to mature. Subsequently, when moisture or fertilization becomes available to the plant, the inner scales resume growth often causing the bulb to split.

Sun Burn

It occurs when the bulbs are exposed to high temperatures and bright sunlight. Damaged onion bulbs from excessive exposure to sun light often have a bleached appearance when scales that may be soft and wrinkled. Onion leaves are often placed over bulb after they have been harvested and windrowed for curing. This protects the harvested bulbs from direct sunlight.

Scorching

Burning during drying in the field caused by excessive sunshine.

Cracked and Double Bulbs

These are caused by sharp variations in soil humidity.

Garlic

A typical disorder, sprouting of bulbs in fields has been reported in some parts of India especially in Saurashtra (India), mostly towards the start of maturity stage of bulbs. This phenomena is noticed when there has been winter rains or excessive moisture and supply of nitrogen. However, this disorder is not of a permanent nature and the loss due to this disorder is not alarming that is not more than 0.5 per cent. It has further been noticed that this disorder is more pronounced in white cultivars than in pink or purple cultivars (Pandey and Singh, 1993).

A disorder known as waxy breakdown, in which the clove shrinks and becomes amber, translucent, and waxy or sticky to the touch, occurs in stored garlic and sometimes in the field. It has been associated with sunscald in the field but usually associated with poor ventilation and lack of oxygen during storage and transport.

Strawberry

Albinism

It is a physiological disorder of straw berry fruit, occurring primarily at the ripening. It has attained an alarming magnitude in strawberry growing countries of the world. Fruits suffering from this disorder appear bloated and develop white or pink areas on their surface with pale coloured pulp. The fruits have poor flavor which do not ripen uniformally and show waxy appearance. Affected fruits are liable to severe damage during harvesting and become susceptible to fruit rot during storage, with the result such fruits fetch low price in the market. Many reasons have been attributed with albinism; having excessive growth owing to heavy nitrogen fertilization. A study conducted at IARI, New Delhi during 2000–2001 on different varieties or straw berry by Sharma and Sharma (2004) revealed that variety Sweet Charlic had semi vigorous growth and had low incidence of albinism. The study further revealed that albinio fruits of all the varieties were almost without colour but the size and quality (TSS) were not significantly different from the normal fruits. So albinio fruits might not have attractive colour but were not inferior in quality. According to Greathead *et al.* (1966) albinism is a disorder appearing in some cultivars of strawberry, were fruits are bitter, acid and have little or no red colour, which appear with excessive nitrogen and potassium application or low light intensity and seems to be related to greatly reduced sucrose transport and calcium deficiency. Possibly, the biochemical pathway resulting in this disorder may be similar to that in mutant genotypes.

Lower dry matter content of albino fruit than normal fruit indicated that albino fruit are more hydrated, which may be due to increased competition between leaves and fruits for different nutrients during the period of excessive vegetative growth, because luxuriant growth favours albinism in straw berry (Sharma and Sharma, 2003b). Among nutrients, potassium was notably higher and calcium was lower in albino fruit than normal fruit, which might have resulted in higher ratios for N/ca and K/ca for albino fruit, there by giving some indications that albino fruit are physiologically riper and more senescent than normal fruit, though they have, poor colour development (Marcelle, 1984). This is further supported by the observation that LOX activity was greater in albino than normal fruit. Lieten and Marcelle, in 1993 have also observed lower LOX for albino than normal (healthy fruit). Cutting *et al.*, 1992, have also shown that calcium content of avocado decreased and K/ca ratio increased is not directly involved in albinism in strawberry. However, LOX may be involved in senescence due to which, it might have increased in albino fruit (Lieten and Marcelle,1993).

Strawberry Tip Burn

This is caused by excess moisture, and the tips of the young leaves, are deformed and die "burnt".

Asparagus

It has generally been noticed that yield reduction commonly occurs in old asparagus. Much of this reduction can be attributed to stand reduction. Replanting new seedlings into these fields is often unsuccessful. Evidence showed that substances produced by asparagus plant are both autotoxin and allelopathic (Hartung *et al.*, 1989). These include several cinnamic acids (caffeic, ferulic, methylenedioxy cinnamic acid), which have been isolated from asparagus roots and are known to inhibit germination of other crops (Hartung *et al.*, 1989) and asparagus Miller, *et al.*, 1991). Recently, an interaction between these autotoxins or allelopathic compounds and fusarium has been observed, where these chemicals have direct physiological and biochemical effect on asparagus that predisposes them to fusarium diseases (Hartung, *et al.*, 1989). Thus replant problems in old asparagus fields occur from exposure of new crown or seedlings to fusarium and autotoxin present in the soil.

Tip Rust or False Rust

This gives rise to long reddish streaks along the stems and is the result of a metabolic disorders in cold moist weather.

Die Back of the Young Shoots

This is attributed to several causes, such as boron deficiency or insufficient water absorption.

Autum Shoot Production

When conditions are favourable in the autum, shoots may be produced, but this is determintal to the next springs crop, as these shoots are using upon the reserves laid down for next springs shoots. To prevent this, avoid watering in this period and apply fertilizer in the winter.

Mushroom

Ryall and Lipton (1972) reported that principal post harvest disorders of cultivated mushrooms result from too slow handling of the product and high temperatures. On the other hand relatively high humidity causes unattractive elongation of the stem and a slim surface unless the temperature is low. Whereas slow handling of mushrooms at high temperature results in brown discolouration of the cap and wilting of the entire structure. Inadequate oxygen supply and increased atmospheric CO_2 (1-2 per cent v/v) can lead to proliferation of the fruit bodies. Similarly, inadequate aeration coupled with lack of suitable illumination is known to cause the development of long slender stalks in the fruit bodies of *Pleurotus flabellatus* (Rajarathnan and Bano,1987)

Distortion

Fluctuations in environmental conditions may induce many forms of distortions *viz.* hollow stem, split strips, swollen stems, hard gills and misshapen mushrooms.

Water Logging

Water soaked areas appear on the stems and if these mushrooms are squeezed water oozes out. In severe cases there is spontaneous release of large quantities of clear or coloured liquid from the mature mushroom, which subsequently collapses.

Pinning Disorders

There are number of pinning disorders, with identical

causes, these disorders are more common and devastating, as they affect both the management of harvest and the yield. They include mass pinning, clumping, stoma pin death and rose-comb.

Chicory

Kranhnstover *et al.* (1997) described 20 physiological disorders of Witloof chicory that may occur during the forcing process. They therefore, suggested that preventive measures during root production, storage and harvest and during all post harvest phases are extremely important. The disorders likely to occur are loose, unfilled or open heads; rosetted, shortened heads; black-blue leaf tint; blind root (axial buds with no main bud); brown flecks on the central axis; brown leaf edges; low temperature damage (oval red-brown areas on outside leaf surfaces) and black (necrotic) flecks on leaves. Den Outer in 1989 described another disorder *viz.* internal browning- a discolouration of the central part of the stem and described it to be a calcium-related disorder, similar to tip burn of lettuce.

Radish

Radishes are also susceptible to freezing injury. Ryall and Lipton (1972) stated that freezing followed by thawing causes the injured tissue to appear translucent. Where in severe cases, root softens, loose moisture rapidly, and shrivel. While as Parsons and Day in 1970, reported that in red radishes the pigment oozes out of the roots with moisture, leaving them yellowish and bleached.

Turnip

Whiptail: It is caused due to deficiency of molybdenum and is more common in acid soils. Young leaves become narrow, cupped showing chlorotic, mottling especially around the margins, develop deep patches, which ultimately affect the root growth. The affected plants are removed from the field during thinning operation. The appearance of this disorder can be controlled by liming the soil and bringing the pH to 6.5. Apply 1.2kg per hectare of sodium or ammonium molybdate.

Hollow Root

This effect may be due to several causes, such as excess ripening, frost or imbalances in soil moisture.

Split Root

This malady occurs when the soil texture is not smooth enough.

Watermelon

Blossom-end Rot

B.E.Rot is a very common disorder in watermelon. The affected fruits are misshapen with brown, leathery, rotten lesions on the blossom-end. This disorder is most prevalent during and following extended dry periods and is caused by insufficient Calcium in the fruit and by the faulty nutrition associated with irregular moisture supply and high temperature.

The smooth, leathery, firm, dark green or brown delimited areas, about 2.5 to 7.5 cm in diameter develop around the point of blossom attachment. This disorder is thought to be caused by faulty nutrition associated with irregular moisture supply and high temperature. This problem can be reduced to a great extent by liming the soil with dolomitic lime before planting and applying well –

timed irrigations to alleviate prolonged drought periods. Calcium nitratre can also reduce this malady.

Poor flavour and lack of sweetness in the fruit may be due to poor soil fertility that is low in potassium, magnesium or boron and cool temperatures. Other causes could be wet weather, poorly adapted cultivars, loss of leaves from disease or harvesting melons before they are ripe. Poor pollination and fruit set is caused by wet, cool weather, insufficient bee pollinators and excessive vegetative growth as a result of close spacing. Whereas heavy rains cause splitting of fruits.

Muskmelon

Scalding

White patches form as the result of excessively strong sunshine and high temperatures.

Splits in the Melon

This is due to water imbalances.

Pumpkin

Poor fruit set is commonly caused by plants with inadequate spacing between plants. Pumpkin requires bees for pollination, similar to other members of the cucurbit family. Pollinating insects are required to transfer pollen from male to female flowers for fruit set to occur. Excessive nitrogen application may also prevent development of fruits.

Cucumber

Bitter tasting cucumbers can result from moisture stress,

temperature, inappropriate soil, or hereditary. Bitter taste usually occurs during the hotter part of the summer or later in the growing season. Cucumber are only grown on well drained soil with a pH between 6.0–6.5 to avoid bitter tasting fruit. A mulch that conserve moisture and keeps roots cool during hot, dry weather tends to reduce the occurrence of better tasting fruit.

Spinach

The production of a premature flower stalk (bolting) in spinach and ensuing seed production can render the plant unmarketable. This can be overcome when spinach is grown under long days and warm conditions. Any deficits in soil moisture can intensify the effect of heat on spinach grown.

Artichoke

Artichoke Frost

This causes blackish patches to form on the bracts around the heads, which wilt.

Scalding

The young plants die, due to watering during the hours of day light.

Sweet Potato

Growth Cracks

Some varieties in sweet potato have a tendency to crack at maturity, which has been attributed to either excess or shortage of moisture. This disorder can also occur if the harvesting is delayed for few days. The judicious use of potassium fertilizers at sowing will check cracks.

Taro

Metsubre

It is a nutritional disorder of taro, and is supposed to be due to calcium deficiency. The defective corms have smooth or concave top, slightly brownish in colour and are of varying sizes. Application of calcium will control this disorder.

Endive

Brown Heart

It is a disorder that affects endive and escarole (Moline and Lipton,1987). It manifests as marginal browning of immature leaves within the plant. There is no organism is involved with the disorder, rather it appears to be similar to tip burn of lettuce in that it is also related to poor calcium distribution in rapidly growing tissues. Maynard *et al.* (1962) were of the view that it is possible to control the disorder with the foliar spray of calcium salt solution twice weekly as endive types have relative open heads. The spray of calcium salt solution is likely to reduce the incidence of brown heart about eight fold. The heads with this malady may be subject to bacterial soft rot as a secondary problem during the marketing process, therefore, maintaining the temperature close to 0°C can minimize the decay.

Maize

Incomplete pollination results in Kernel skips on the ear. Poor pollination can result when air temperatures are above 35.52°C, when plants are exposed to hot winds or when plants are under moisture stress.

Chapter 6

Abiotic Disorders of Vegetable Crops

Environmental factors and nutrients play an important role in the health of the plant. It is the adequate supply of nutrients, pollution free soil environment and optimum temperature and moisture that favour normal growth of the plants. Therefore, any deviation from these conditions result in the expression of different disorders of various magnitudes. It is both the deficiency or excess of any of the nutrient element *viz.* heavy metals, soluble toxic salts in the irrigation water, toxic gaseous pollution in the air, unsuitable prevalent temperature, moisture and soil pH have direct effect on plant growth. Most of the vegetable crops are sensitive to the adverse conditions and depict different symptoms. Deviation from optimum levels of temperature and moisture results abnormal pattern of growth. Reduced moisture and increase in transpiration results in tomato fruits to develop water soaked spots, which ultimately become sunken, leathery and dark coloured, while at high temperature during rainy season tomato fruits exposed to sun develop concentric cracks. Similarly, premature initiation of floral buds " riceyness" in cauliflower is due to temperature fluctuations. Soil temperature significantly affects the growth of cucumber, brinjal, peas, radish and beans (Wilcox and Pfeiffer,1990). Intensity of light affects senescence and tuberisation in potato (Charles *et al.*, 1992). Sensitivity to chilling depends upon the growing stage of plant. Similarly seedlings are more sensitive than the advanced stages of growth (Lyons, 1973).

Air pollution resulting from sulphur dioxide and ozone has detrimental effect on vegetable growth. The phytotoxicity of ozone has marked effect on secondary metabolism in potato, tomato and other plants. Some species of beans and potato develop brown or black stipples on leaves, while white flecks or bleaching occurs on foliage of onion, tomato and cucumber. Leone in 1977, reported that slivering or glazing on the lower leaf surface of leafy vegetables is caused by the photochemical pollutant known as peroxyacetyl nitrate (PAN)

Low Temperature Disorders

The overall effect is that an imbalance in metabolism is created, and if it becomes serious enough to result in an essential substrate not being provided, or toxic products being accumulated, the cells will cease to function properly and will eventually lose their integrity and structure. These collapsed cells manifest themselves as areas of brown tissue in the produce. Metabolism disturbances occurring at reduced temperature are generally divided into two main groups: chilling injury and the physiological disorders.

Usually storage of produce at low temperature is beneficial, because the rate of respiration and metabolism is reduced. Lower storage temperatures do not suppress all aspects of metabolism. Some reactions are sensitive to low temperature and cease completely below a critical temperature. Decreasing temperature does not necessarily reduce the activity of other systems to the same extent as it does respiration. For these systems, this leads to an accumulation of reaction products and possibly a shortage of reactants, while the conserve occur with labile systems. The overall effect is that an imbalance is created and if it becomes serious enough to result in an essential substract not being provided, or toxic products being accumulated, the cells will cease to function properly and will eventually lose their integrity and structure. These collapsed cells manifest themselves as areas of brown tissue in the produce. The metabolic disturbances occurring at reduced temperature are generally divided into two main groups:

Chilling Injury Physiological Disorder

Chilling injury

Chilling injury is a disorder which has long been observed in plant tissues, especially those of tropical or subtropical origin. It results from the exposure of susceptible tissues to temperatures below 15°C, although the critical temperature at which chilling injury symptoms are produced varies from vegetable to vegetable. Chilling injury is a separate phenomenon from freezing injury, which results from the freezing of the tissue and formation of ice crystals at temperatures below the freezing point. As susceptibility to chilling injury and its manifestations vary widely among different vegetables and susceptibility to this disorder means that the lowest safe temperature for these vegetable crops will be well above the lowest non-freezing temperatures. Therefore, a clear distinction can be made between the causes of chilling and freezing injury. Carrots are susceptible to freezing injury at temperatures below freezing (1.5°C). Chilling injury occurs in sensitive plants at temperatures of about 0°C for sweet potato to 13° C for banana, with many fruits and vegetables in the 4°C to 6°C range (Pantastico,1975). This is a complex disorder with a broad spectrum of symptoms ranging from large, unsightly, sunken pits to surface dulling, darkening, or discolouration to water areas and internal browning, water soaking and loss of flavour or off- flavours. Products of most tropical and some subtropical plants show visual and internal manifestations of the disorder which often occur weeks or months earlier as in the case of banana. Exposure to adverse conditions need be only a few hours as with ripe bananas, or a period from 10 days to several weeks or months. Chilling injury is a physiological disorder which affects plants at all stages of development, so it is not surprising to find the seeds and seedlings of crops of tropical or subtropical origin as severely affected by exposure to temperatures between 0°C and 10°C to 15°C. On the other hand, seeds and seedlings of crops of temperate origin are not injured by these temperatures.

Chilling injuries in plants is recognized as abnormal development after exposure for more or less prolonged periods to temperatures below about 10 –13°C. Plants of tropical or subtropical origin are generally assumed to be susceptible to chilling and abnormal development following a chilling exposure might occur on any part of such a plant and can be expressed in a wide array of deleterious systems.

A common physical symptom is pitting of the skin, which usually occurs due to the collapse of the cells beneath the surface and the pits often become discoloured. The other symptom is the high water loss which further accentuates the extent of pitting. The other most important symptom is the browning of the flesh tissues, which first occurs around the vascular strands in fruit, which may result from the action of the enzyme polyphenol oxidase on phenolic compounds released from the vacuole after chilling. Water-logging is also often observed.

Chilling injury differs from freezing injury as it occurs at temperatures which are low but much above the freezing point of tissues, freezing injury resulting in the formation of ice in intracellular spaces which leads to the death of cells. The visible chilling injury symptoms occur first in the distal end of fruit while the proximal end remains apparently normal. Symptoms become more severe and eventually extend to the proximal end as exposure time to chilling temperature increase. This observation suggests that there is a differential susceptibility to chilling within individual fruit (Singh, *et al.*, 2006).

Symptoms of chilling injury normally occurs when the produce is subjected to low temperature, but sometimes may also appear when the produce is removed from the chilling temperature to higher temperature, which results in its rapid deterioration. The chilling injury of the produce may result in the release of the metabolites such as amino acids. Sugars and mineral salts from cells which together with the degradation of cell structure provide an excellent structure for the growth of pathogens, especially fungi. Rapid intake of free water during imbibition causes a disruption of cellular membranes, increasing leakage from the tissue and causing cell death. For this reason there is increase in rotting in tropical produce after low temperature storage. Chilling often results in the development of off–flavours or odors. Thus, it can be concluded that the complex array of symptoms suggest that there are several factors operative in the development of chilling injury. Thus different vegetables grown in different areas may behave differently.

Symptoms of chilling injury to plants are diverse for example surface lesions, water soaking of tissues, internal discolouration (browning), break down of tissue, failure of fruits to ripen in the expected pattern, accelerated rate of senescence, increased susceptibility to decay, shortened storage or shelf life, compositional changes related to consumer acceptance and loss of normal growth capacity. Most of these symptoms are not unique to chilling.

Tomato is one of the most frost-sensitive of vegetable. It shows little adaptation to hardening process which is characterized in cabbage. A mature tomato crop is exposed to near-freezing temperatures, the accumulation of the sugars may lower the freezing point slightly and thickening of the cuticle may have some protective value. There is little difference in varieties in adaptation to low temperatures. Tomato fruits may be under cooled below their freezing point and remain unfrozen for a time if not disturbed. The average freezing point of most of the tomato varieties was found to be –0.86°C (Harvey and Wright,1922).

Chilling injury in tomato is characterized by a range of symptoms including an increased rate of ripening, extensive green patches on otherwise red fruit, an uneven surface due to collapse of cells and the production of excessively soft fruit (Hobson, 1987). More important, from texture perspective, is the observation that tomatoes with slight chilling injury had developed a mealy texture (Jackman, *et al.*, 1992a). Mealiness was characterized by the absence of expressed fluid during compression of disks of pericarp tissue and may be associated with elevated activity of cell wall degrading enzyme ß-glucosidase (Jackson, *et al.*, 1992a), and higher levels of pectimethylesterase (Marangoni *et al.*, 1995).

Symptoms

At early stages of chilling the incipient changes are reversible, but once fully effected they are irreversible and lead to considerable metabolic imbalance and chilling injury reflect a wide range of metabolic dysfunctions in chilling.

1. Chilling impaired the photoplasmic streaming.
2. It reduces respiration, photosynthesis and protein synthesis
3. It alters the patterns of protein synthesis.
4. Chilling affects the plants at all the stages of development starting from initial growth to ripening.
5. Chilling reduces the ethylene production which is responsible for fruit ripening.

Symptoms of chilling temperature to plants are diverse for example; surface lesions, water soaking of tissues, internal discolouration (browning), break down of tissue, failure of fruits to ripe in the expected pattern, accelerated rate of senescence, increased susceptibility to decay, shortened storage and shelf, compositional changes related to consumer acceptances and loss of normal growth capacity. Most of these symptoms are not unique to chilling.

Chilling injuries to temperate–origin crops include;

- Surface symptoms- For example Chinese cabbage may develop browning of the mid rib.
- Internal discolouration and tissue break down. Certain potato cultivars can develop mahogany browning of their flesh.
- Compositional changes; Beet may become abnormally hard and develop earthly flavour.
- Abnormal ripening
- Loss of growth capacity
- High perishability; Broccoli store best at 0°C, yet off odors under restricted ventilations were more pronounced at 0°C than 2.5 or 5°C.

Freezing Stress

In nature, plants cool slowly and ice always forms in the intercellular spaces or on the surface of the tissue, resulting in extra cellular freezing. When plant cells are cooled rapidly or inoculated with ice at a considerably super cooled stage, cells under go intracellular freezing (ice forms in side the cell). Intracellular freezing can be further classified into flash and non-flash types (Asahina, 1978). Flash type is a sudden freezing characterized by an instantaneous darkening of the entire cell. During non flash freezing, ice growth is visible in the cell (Asahina,1956). A high rate of cooling is favourable to flash freezing. Non-flash freezing can occur in super cooled cells. Although cooling rates have a major influence on the type of intercellular freezing that occurs, the character of the cell itself may also bear some relation to the type of intercellular freezing. Cells from frost sensitive plants like tomato and melon fruits, usually freeze in a flash type manner even when cooled slowly. Cells from non-acclimated plants capable of cold-acclimation, such as cabbage and spinach, exhibit non-flash freezing at moderate cooling rates, *e.g.*, 4°C/min. Ice growth in the cells can be seen in a period of few seconds. Cells frozen in this fashion appear clear, microscopically.

The protoplasts become dehydrated during extra cellular freezing as the water moves from the protoplasts to the extra cellular ice. As a result, cells under go dehydration and contraction. The amount of ice formed is a function of temperature. Olien in 1967 termed this type of freezing " equilibrium freezing"- in frost sensitive plants; the amount of ice formed may not be a function of temperature drop and this is termed "non-equilibrium freezing." During extra cellular freezing, water moves from the cell, mainly from the vacuole, to the intercellular spaces, forming ice; this process results in cell dehydration. As freezing proceed, ice growth causes cellular contraction. Thus, during freezing, a cell can experience three types of stress, freezing induced dehydration stress; osmatic stress due to removal of water from the vacuole; and mechanical stress caused by ice growth and cell contraction.

Extra cellular freeze can fairly injure the cells when they are cooled beyond a tolerable limit of low temperature.

When plants are subjected to subzero temperatures the cell water super cools if there are no sites of ice nucleation. A high concentration of cell sap may induce a few more degrees of super cooling; thus the tissue could tolerate a greater freeze stress by avoiding ice formation.

On freezing vines become dark, wither and desiccate, where fruits on freezing become soft and water-soaked or dull coloured and decay after the activity of the secondary organisms.

Lettuce freezes at –6±0.2°C therefore, care must be taken that it is not exposed to low temperature for long enough to freeze the leaves. Carrots are susceptible to freezing injury at temperatures below freezing (-1.5°C).Radishes are also susceptible to freezing. Ryall, and Lipton (1972) have stated that freezing followed by thawing causes the injured tissue to appear translucent. In severe cases, rots soften, lose moisture rapidly and shrivel. Parsons and Day in 1970 reported that in red radishes the pigment oozes out of the roots with the moisture leaving them yellowish and bleached. In case of lettuce freezing injury is characterized by blistering and subsequent dry and darkening of the epidermis of the outer leaves when it freezes in the field or during storage. Since lettuce freezes at –6±0.2°C (Ellison *et al.*, 1981), care must be taken that the lettuce is never exposed to lower temperatures long enough to freeze the leaves. Beets are susceptible to freezing injury between–1.0 and –1.5°C, hence should not be held where they may cool below their freezing point for more than a few hours (Ryall and Lipton, 1972). In case of turnips chilling injury is indicated by surface pitting, shriveling and decay.

In tomato, the first visible sign of freezing to death of fruit tissue is the appearance of small or extensive water-soaked spots or areas. Fruits do not freeze so readily as the foliage, and those in contact with the ground may be less injured than those free from the soil. Temperatures which have not been sufficiently low to cause immediate injury may be detrimental. Tomatoes if held for 4 days at 0°C showed no injury, and ripening proceeded normally when exposed to room temperatures, but longer exposures (8 days) lead to break down and decay (Diehl, 1924). A yellow blotching of fruit several days following a severe frost has been noted, and apparently uninjured fruits picked following a frost has been reported to decay more than normal stock (Wright *et al.*, 1931). Chilling of green tomatoes (even at –3.8°C for 18 to 21 hours; 5 to 8 days at 0°C; or 11 to 15 days at 4.4°C).

It does not prevent normal ripening when removed later to higher temperatures, but ripening may be delayed.

In potato fields low temperature injury occurs when there is frost or snowfall. There are no external symptoms, but internal necrosis is the main symptom. The tissues of the stem end of the tuber are in general more sensitive to freezing injury than those of the eye buds, and the differential vascular cells such as tracheae, sieve tubes and tracheids are more susceptible than parenchymatous cells. There are three types of necrosis *viz.* ring necrosis, net necrosis and the blotch necrosis. In ring necrosis there is discolouration in the region of the vascular ring and more commonly at the stem end. While in net necrosis finer vascular elements of the inner phloem, scattered throughout the tuber within the vascular ring are blackened. In blotch necrosis there are irregular areas of different size showing opaque grey to black discolouration. These three necrosis occur according to the time for which the tubers were exposed to low temperature. The blotch necrosis is the result of maximum exposure.

The symptoms of chilling injury in asparagus are flaccidity and a dull, dark, grey-green aspect of the tips and some times of the zone of elongation below it. Freezing injury is expressed as a darkening and flaccidity of the tips and as a water soaking in the lower portions. Although the symptoms in the tips resemble those of chilling injury; the mushiness that develops upon thawing allows pinpointing the problem.

Chilling Injury Symptoms of Some Vegetable Crops

Produce	*Approx. Lowest Safe Storage Tem. (ºC)*	*Symptoms*
Brinjal	7	Surface scald
Melon	7–10'	Pitting, surface rots
Tomato	10–12'	Pitting, Alternaria rots

A range of temperature indicates variability between cultivars in their susceptibility to chilling injury.

The damage due to chilling injury can be controlled provided the critical temperature for its development in a particular vegetable is determined and the produce is not exposed to the temperatures below that critical temperature. Chilling and freezing tolerance have been associated with increased ABA levels. Similar increases in ABA levels have been observed with triazole; induced chilling tolerance in banana.

Therefore, the ability of triazole compounds to alter the ABA balance may play a role in protection from low temperature stress.

Lightning Injury

The outdoor tomatoes occasionally are affected during electrical storms. When a strong electrical potential is discharged in tomato field, the plants in the vicinity are either killed or injured in a circular or elliptical area of up to 18.3 m. The damage is severe in the center of the area. The first symptom is the drooping of the leaves at the extremities, while in severe cases there is progressively wilting and death of the plants. In less severe cases the plants recover after the initial wilting. The stem, branches and petioles show various degrees of progressive shrinkage due to injury to and collapse of the pith. The cortical tissue shows various amounts of bronzing or bleaching. The charge passing into immature fruits from the stem causes varied effects. Browning in the fleshy tissue may extend to the periphery. On the outside of the fruit, lesions appear, ranging from small necrotic spots to large, irregular, blister-like areas which shrivel and desiccate (Whipple, 1941).

High Temperature Injury

In many parts of the world high temperature causes the heads of broccoli to be rough, with un-even bud sizes. High temperatures can induce numerous physiological and biochemical effect in plants, including protein denaturation, enzyme in activation, altered metabolic rates, membrane damage, reduced chloroplast biochemical activity (Blum,1988). These effects of heat stress are often confounded with those of water stress, since high temperatures are accompanied by increase transpiration rate and dehydration (Kramer,1980). The close relationship between these two stresses is supported by a study that found increase thermal tolerance in triazole- treated wheat plants that had been pre-exposed to water stress. The principal types of heat injury are (1) retarded growth and undersize or failure to mature the flowers and fruit; (2) localized killing of tissues or a sunburn or sun scald of leaves, flowers or fruits; (3) localized killing of stem tissues or formation of heat cankers; (4) defoliation or premature shedding of leaves; (5) premature ripening of fruits; and (6) death of the plant as the result of a general heat necrosis. High degrees of heat are frequently accompanied by intense sunshine and extremes of drought which intensify the injury. Death of cells from high temperature results when there is an irreparable destruction of the molecular structure of the cytoplasmic body.

Tomato Magnesium Deficiency
(*Source*: Cornell University, Ithaca, New York)

Blossom End Rot of Tomato
(*Source*: T.A. Zitter, Cornell University, Ithaca, New York)

Lettuce Drop
(*Source*: T.A. Zitter, Cornell University, Ithaca, New York)

Lettuce Big Vein
(*Source*: T.A. Zitter, Cornell University, Ithaca, New York)

Sulphur Deficiency
in Cauliflower

Sulphur Deficiency
in Pea

Nickle Toxicity in Fenugreek Plants

Potassium Deficiency in Potato Plant

Potassium Deficiency in Turnip Leaf

Cabbage with S (L) and without S (R)

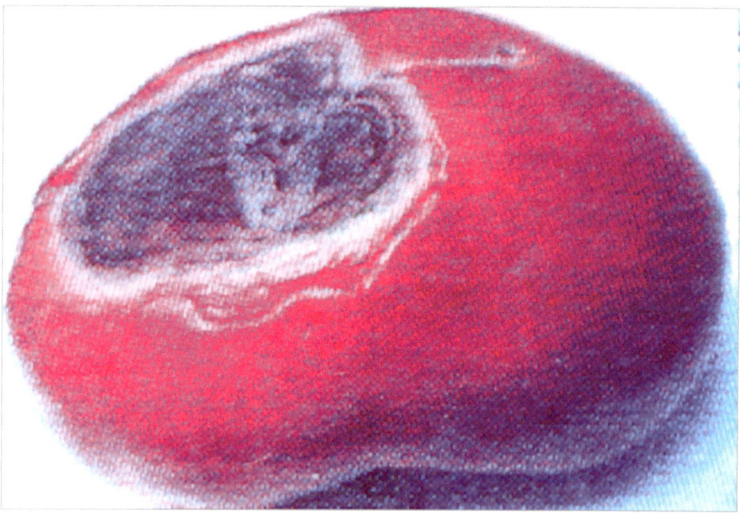

Calcium Deficiency in Tomato

Chapter 7

Disorders in Vegetables Attributed to Controlled Atmosphere (CA)

Numerous disorders have been attributed to CA, without much evidence as to their nature, except for aerobic respiration damage. Lipton (1976b) has observed that damage due to high CO_2 or low O_2 invariably occurs on those tissues lacking chlorophyll, such as leaf midribs and never on tissues having chlorophyll. He cites the following examples: Cauliflower curds are injured in a C.A of 5 per cent CO_2 and 2 per cent O_2, while broccoli keeps very well in 10 per cent CO_2 and will tolerate O_2 as low as 0.25 per cent; yet these are plants of the same species. Crisped lettuce with much white tissue is damaged by as little as 2 per cent CO_2, while Romaine lettuce, with mostly all green leaves, is not damaged in 12 per cent CO_2. differences in growing locations also affect the susceptibility of a cultivar to CO_2 damage. When held in 10 per cent CO_2 'Climax' lettuce grown in the Salinas valley suffered brown strains (BNS) damage, but that grown in desert areas had no damage, Lipton suggested that CO_2 damage to white or yellow tissue might be related to soluble solids content (SSC). He has tested this hypothesis (but not yet conclusively) by harvesting lettuce at 0700 and again at 14hrs. Some validity to SSC theory is suggested because lettuce harvested in the morning suffered greater BNS damage than that harvested in the afternoon, after a period of photosynthetic activity.

Morris and Kader (1977b) reported that mature green tomatoes when held in 5 per cent CO_2 for more than 7 days at 20°C or 10 days at 12°C, developed surface blemishes, softening, and ripening unevenly. Bell peppers stored at 10 per cent CO_2 at 12.5 per cent for 7 days had calyx discolouration, softening and intervienal browning.

Because of the numerous kinds of plant tissues encountered in such a broad category as vegetable species, it is difficult to develop a set of criteria that positively signifies high CO_2 damage or low O_2 damage. However, the observations of Lipton concerning the greater susceptibility of achlorophyllous than chlorophyllous tissues to damage need further investigation.

Ozone

It is extremely phototoxic. It oxidizes plant surfaces and tissues and affects many important physiological process. It is particularly injurious to membranes and notably inhibits photosynthesis and biomass accumulation, suppresses phloem loading, and reduces carbon allocation to the roots. These actions have resulted in ozone being rated as the pollutant of major concern in crop production in forest tree growth (Table 19).

Role of Nutrient Elements in Vegetable Crops

Many economic crops receive generous application of natural and chemical fertilizers, but as a group none are usually fertilized at higher rates than vegetables, because they are mostly short term crops, where production is associated with a high and continuous level of fertility throughout their growth period. Even temporary restriction in nutrient availability may have catastrophic effect on the economic value of the crop by delayed maturity, lower yields or impaired quality.

Vegetables grown for their succulent leaves, petioles or stems must be continually grown under favourable nutritional regimes to insure dark green colour, succulence and high yields. Accordingly, lettuce, celery, parsley and similar crops are liberally fertilized throughout their growth period. This is accomplished by a combination of pre-planting, planting and side dressing applications of fertilizer. The practice of multiple applications combined with liberal rates of fertilization has several actual or potential disadvantages, including loss of nitrogen through leaching, di-nitrification or volatilization, danger of crop damage; and high labour and energy requirements.

Soil nutrients are the basic source of the farm nutrient supply. The part of them are utilized by crops *i.e.* the easily available portion (water soluble, exchangeable) as well as the easily mobilization fraction. The mobilization of nutrients from (mineral and organic) slowly available sources can be enhanced to a certain extent by activating soil life (in general by organic manure or by special bio-fertilizers), by crop varieties with strongly mobilizing capacity, by better accessibility of nutrients after structure, improvement, by deepening of the plough layer or by fallow periods.

The best use of nutrients for crop growth can be obtained on the basis of a high soil fertility level. Soil fertility is a complex term (not very precise, but very useful even in its vague form) which include many complex components: soil depth, texture and structure (pore space for supply of oxygen and water), soil reaction, organic matter content and composition, activity of soil organisms, nutrient content, storage capacity for nutrients, respectively, absence of detrimental or toxic substances. The result of optimum combination of these factors is a high soil fertility that means a high crop (vegetable) production potentials

The range of soil is so wide that some of the soils are having exceptionally low or excess amounts of certain nutrients. The parental material and soil forming processes often determines the nutritional stress problem in soils and many nutrient deficiencies or toxicities can be often predicted from these. Moreover, the long history of cropping the soil, which was once sufficient in nutrients, become depleted in nutrients and hence the nutrient deficiencies are wide spread.

The term 'plant nutrient' generally refers only to mineral nutrient and like other plants vegetables also require all 17 elements for growth and development. C, H and O, assume to be plentiful and mainly require for photosynthesis are taken up by the plant from air, air space of the soil and water and hence these are called non-mineral nutrients.

The essential elements are assimilated in to plant through absorption by root or other plant parts as ions from the soil and hence are described as mineral nutrients and based on their requirements by plant these are classified as macro-nutrients (N,P.K. Ca. S and Mg), and micro- nutrients (Fe, Mn Zn, Cu, B, Mo and Cl), in addition nickel (Ni), Cobalt (Co) and sometimes Al are also beneficial.

Nutrient Deficiencies

Adequate and balanced nutrition is essential for sustaining crop productivity. The causes of nutrition deficiency may be either direct or indirect, when the soil is inherently deficient in a plant nutrient (s), crop production will be directly affected. The threshold concentration below which the nutrition and hence growth and production of the crop are affected depends not only on the nutrient element but also on crop requirements. In other words, the concentration of a nutrient in the soil considered adequate for a crop may be inadequate for sustaining growth and productivity of another. For this reason, the soil critical level of a nutrient defined for a crop may not apply to another crop species. In contrast to inherent infertility of the soil, artificial scarcity of a nutrient (s) may also develop in certain situations. In this case it is not the absolute deficiency of the nutrient that effects the growth of the crop but under-utilization of the soil nutrient resources by the crop that is affecting the growth. Adverse soil conditions such as soil composition, presence of sub soil hard pan etc; can lead to restricted root growth and reduced exploration of soil nutrient. The important causative factors of nutrient deficiencies are *viz.* soil type, soil pH, nutrient requirements etc.

Management of Mineral Disorder Symptoms

Major Causes of Mineral Disorders

Plants require all the essential nutrients in balanced proportions and deviation from this may result in mineral disorders. This disorder may be due to deficiency or toxicity of a particular nutrient or multi nutrient. When two or more elements are deficient or toxic simultaneously, the composite picture of symptoms may resemble no single known symptoms. Generally, nutrient deficiency in the plant occurs when a nutrient is insufficient in the growth medium and or can not be absorbed and assimilated by the plants due to unfavourable environmental conditions. Nutrient disorders limit crop production in all types of soil, throughout world. Plant grows in an environment facing different climatic conditions and soil types and hence there are several factors causing these nutritional deficiencies.

1. Continuous withdrawal and inadequate supply of nutrients in the soil
2. Edaphic factors (soil, water, temperature and environmental) preventing absorption of nutrients by plants
3. Changes in soil physico-chemical conditions such as pH and EC
4. Imbalance use of fertilizers
5. Induced deficiency
6. Interaction between minerals during uptake
7. Antagonism
8. Use of intensive nutrient (response) requiring crops and nutrient inefficient crops
9. Biotic (Disease and pest) factors.

These factors are inter-related and interaction between them is very complex for the absorption and utilization of nutrients by plants. Therefore, early detection of nutritional deficiency stress is important as these might extend to the entire plant if relief of stress is not employed and continuous shortage of a nutrient or nutrients might cause plant death. Some of these factors are discussed here in detail.

Acid Soils

Both acid and alkali soils cause nutritional stresses in crops. Acid soils having pH less than 5.5 are deficient in Ca, P, and Mg, but abundant in Al, Mn and Fe causing major nutritional problems for vegetable crops. Such soils have degree of saturation of CEC below 25 and therefore, problem with acid soils is due to toxicities of Al, Fe and Mn. Besides such soils have low water holding capacity, susceptible to crushing erosion and compaction thus making them low productive. Therefore, are more detrimental to crop production.

Calcareous and Alkaline Soils

Soils having pH value above 7 are calcareous and alkaline. Calcareous soils have high amounts of lime and lime stone potential below ground and the high HCO_3 is associated with increased pH or CO_2 concentration, which reduces the solubility of Fe and Mn causing their deficiencies. Generally the alkaline soils have abundant Ca, Mg and K but are deficient in S and toxicities of B may occur. The deep ploughing which brings up carbonate of higher solubility on to the surface layers and increases the soil pH should be avoided. The phosphorus availability is low in calcareous soils and added P depends on number of weekly absorbing sites. As the high bicarbonate content directly affects the uptake of P, therefore, its utilization of P depends on plant tolerance to alkalinity.

Light and Temperature

The concentration of elements are influenced by light intensity, its duration and temperature extremes. Where as the light affects the photosynthates produced and alter the ratio of element to dry matter due to dilution effects. Light through photosynthesis provides energy for active uptake of elements, thus enhancing the concentration. But the positive effect is over-ridden by dilution effect. Thus increasing light exposure reduces the N, P and K concentration, but Ca concentration may increase (Jones, *et al.*, 1991). While shading increases the P, K, Al, Ca, Fe and Mn concentrations but decreases Cu, Mg and Zn concentrations in leaves. The adverse effects of excess Na, and K associated with high salinity and low light have been partially corrected by high light by helping to maintain plant cation concentration.

Temperature influences the movement, translocation and utilization of elements by plants, but it is difficult to distinguish the effect of temperature under normal field conditions. There is optimal temperature for plant growth and development, which may vary from species to species and deviation from there, definitely cause abnormality in nutrient uptake. It has been studied that the elemental concentration at high temperature is less than that of at low temperature, because high temperature, like light, increases the dry matter production effect. High temperature also enhances respiration, depletes photosynthates and increases transpiration which allows more absorption of elements from soil.

Water

The rainfall increases soil moisture by bringing the levels within the beneficial range between wilting and field capacity, stimulates the plant growth and hence tends to low elemental concentration again due to dilution effects. Though the favourable moisture condition also influences the soil nutrient availability and their movements in plant all these are over-come by fast growth and dry matter accumulation. The humidity influences the rate of transpiration and thus, immediately affects the nutrient content in plants. Water deficit stress results in loss in turgidity and firmness, in addition to these symptoms, there are other undesirable quality changes. Water loss can result in the discolouration of leafy vegetables, loss in aroma and flavour, decline in nutritional value due to losses in mineral nutrients and vitamins, increased susceptibility to other disorders, incidence of pathogens invasion, accelerated rate of senescence.

Flood irrigation (too much of moisture) aggravates the problem of root aeration in vegetables, resulting in abnormal respiration, inhibiting root growth and altering metabolism functions. Because of this plant, becomes chlorotic due to deficiency of N caused by inability of roots to take up N and ineffectiveness of nitrogen-fixing bacteria, deficiency of and due to leaching in coarse textured soil and deficiency of Fe in calcareous soil due to conversion of ferrous to ferric form.

Interactions of Nutrients

Increasing the levels of any macronutrients, up to certain level in the nutrient solution, increases the concentration of that particular element in leaves, stems and seeds and their uptake by plants. But low or excess of any element influences the uptake of other nutrients also that may be synergistic (beneficial) or antagonistic (detrimental) effect, *e.g.* increasing the level of N increases N,P,K, Ca and Mg concentration, while as K decreases Ca and Mg and calcium decreases K and Mg concentrations in plant tissues.

Heat Stress

Temperate plants generally die if exposed to temperatures of 44°C – 50°C, but there are some who can tolerate higher temperatures (Salisbury and Ross,1992). As defined by Levitt (1980), direct (Primary) heat injury in plants, induced by short exposure to extreme temperatures, consists of coagulation of protoplasma, protein denaturation and/or perturbation of membrane integrity. However, indirect (secondary) heat injury results from longer exposures to less extreme temperatures include biochemical lesions and metabolic imbalance.

Plant responses to stressful environmental factors can be part of the mechanisms that permit them to withstand the stress. Acclimation, a process that increases stress tolerance, may occur in response to mild non lethal stress. Many of the changes that appear during acclimation to heat stress are reversible, but if the stress is too great, irreversible changes occur and these can lead to death.

Essential Elements

Sixteen elements are essential for plant growth and reproduction. Three are obtained from air and water- C,H and O; the remaining 13 are usually absorbed by plant roots from growing media. Six of these 13 (N, P, K, Ca, Mg, S) are required by plants in relatively high concentrations and are referred to as macro-elements. The other seven are required by plants in much smaller concentrations and are referred to as micronutrients. These are iron, manganese, zinc, copper, boron, molybdenum and chlorine.

Macronutrients

Nitrogen

Role in Plant

Of the three major plant food elements nitrogen has the most noticeable effect on plants, with effects showing soonest. Nitrogen encourages above ground vegetative growth and gives a dark green colour to leaves. It tends to produce soft, tender growth, a good quality for crops such as lettuce to possess. The tender growth makes the plant better tasting. Nitrogen also seems to regulate the use of other major elements. But too much nitrogen may lower the plants resistance to diseases, weaken the stem because of long soft growth, lower the quality of fruits, causing them to be too soft to transport. It delays maturity or hardness of tissue and thus increase winter damage to plants. Nitrogen is probably the most important nutrient. The deficiency is fairly common, especially on light textured soils, lacking organic mater. It is used in protein and chlorophyll formation and is very effective at stimulating vegetative growth. Most plants prefer to take nitrogen in the form of nitrate (NO_3). Nitrite (NO_2) is toxic to most plants. Nitrogen levels are extremely difficult to measure because (i) it is highly soluble and will be leached out quickly, so readings for fertilizer application are useless (ii) nitrogen is continuously being mineralized (recycled) from soil organic matter. It should therefore, be applied as little as possible as often as possible. For this purpose of fertilizer application, therefore nitrogen levels are estimated based upon previous plant growth, summer rain fall and soil texture.

It is required by plants in higher concentrations than any other element, except K in some instances. It is an important plant nutrient, and is a constituent of proteins, chlorophyll, amino and nucleic acids and is required for the vegetative and reproductive growth, nutrient absorption, photosynthesis and production of assimilates for developing pods in peas and beans. Nitrogen compounds constitute 40 to 50 per cent of dry matter of protoplasm. It also plays an active role in enzyme reactions and energy metabolism. The relative amount of N to carbohydrates in plants reflects the ratio between proteins and stored carbohydrates and thus, the type and quality of growth and flowering. The nitrogen requirement of vegetable crops is much more higher. It is absorbed by plants in the form of nitrate and ammonium ions and as urea. Once inside the plant, nitrate is reduced to NH_4^+ N using energy provided by photosynthesis. The supply of nitrogen is related to carbohydrate utilization. When its supplies are insufficient, carbohydrates will be deposited in vegetative cells, which will in turn cause them to thicken. When its supply is inadequate, and conditions are favourable for growth, proteins are formed from the manufactured carbohydrates. Less carbohydrate is thus deposited in the vegetative portion, more protoplasma is formed, and because protoplasm is highly hydrated, a more succulent plant results. The excessive succulence in some crops may have harmful effect. Excessive fertilization (N) will also reduce the sugar content of sugar beets or in some cases excessive succulence may make a plant more susceptible to disease or insect attack.

Nitrogen can be supplied in the form of organic (or natural) fertilizers, such as castor bean meal, cotton seed meal, fresh meal, dried blood, sewage sludge, or garbage tank-age, but these forms of N are expensive and slowly available. Nitrogen is in the form of proteins in organic materials and these must be biologically degraded to ammonia and/or nitrates before plants can absorb them. Nitrogen is lost from the soil very easily by leaching (washing out). It is very soluble in water and is not held by the soil particles because of the charges of the particles involved. Soil particles have a negative charge, nitrogen also has a negative charge. Since like charges repel, nitrogen particles are not held in the soil. However, organic matter does tend to hold insoluble nitrogen which is released slowly into the soil. Nitrogen is quickly lost from the soil through leaching, especially in sandy soils which lose water faster. It can damage plants if applied in too great an amount.

In summary nitrogen is essential constituent of amino acids, amides, nucleotides and nucleoproteins and is essential to cell division, expansion and therefore growth. It is mobile in plant.

- ☆ Nitrogen moves to younger tissue, so deficiency is first visible in older leaves.

- ☆ Nitrogen is essential for nutrient absorption, vegetative and reproductive growth.
- ☆ It is an important constituent of proteins, chlorophyll, amino and nucleic acid.
- ☆ It is required for photosynthesis and production of assimilates for developing pods in peas & beans.

Deficiency

Nitrogen is very mobile within plants, and deficiency symptoms of this element first occur in older leaves. Protein in older leaves degrade to amino acids (as N becomes limiting) and are translocated to young tissue where they are re-synthesized into proteins. When plant are deficient in nitrogen, they become stunted, yellow in appearance. The first indication of N deficiency is development of a pale green colour in older leaves, which progressively turns to green- yellow, yellow- green, yellow and cream as the deficiency becomes more severe, and young leaves get smaller. This yellowing or chlorosis, usually appear first on the lower leaves: the upper leaves remain green. In case of severe nitrogen shortage, the leaves will turn brown and die. Chlorosis is general over the entire leaf surface and older leaves usually drop before necrosis occurs (Dickey, 1977). If nitrogen is lacking in later stages of development, the stem are short and thin. In some plants chlorotic leaves may develop orange or reddish colours. The leaves fall pre-maturally and shoot and root growth is restricted. Branching, flowering, fruit and seed production all decrease. A deficiency interrupts growth processes, causing stunting, yellowing and reduced dry matter yield.

In potato its deficiency is expressed by light green to yellowish–green colour of the foliage. Its deficiency restricts not only the yield and quality of the produce but reduces the growth of plants and causes the loss of chlorophyll. Both root and top growth is stunted. Nitrogen deficiency in vegetable crops will occur if the leaf total nitrogen content falls below 1.5 percent of the dry weight (Maynard,1979).

Deficiency Symptoms in General

- ☆ Stunted growth
- ☆ Light green to pale yellow leaves starting at the leaf tip, followed by death of the older leaves if the deficiency continues.
- ☆ In severe cases the leaves turn yellowish-red along their veins and die off quickly.
- ☆ Flowering is both delayed and reduced.
- ☆ The plants grow slowly, become stiff and upright in habit, develop short, narrow leaves of a very light colour. Latter the tips of the older leaves die and become bleached and yellow.
- ☆ Chlorosis is general over the entire leaf surface and older leaves usually drop before necrosis occurs (Dickey,1977).

The deficiency of nitrogen (Table 4) limits cell expansion and cell division. Deficiency symptoms include general stunting and yellowing, particularly of the older plant parts. The reduction in plant growth can cause accumulation of sugars and in some species causes the basal tissue to turn purple due to anthocyanin formation. Being highly mobile in plants, the younger leaves and developing organs with strong sink demands such as fruits and seeds, draw heavily on nitrogen in the older or lower leaves. The result of such redistribution when nitrogen uptake is limited in firing (yellowing & senescence) of the lower leaves. The deficiency can be corrected, as signs appear, by application of readily available nitrogenous fertilizer to the soil in the vicinity of the plants.

The visual deficiency symptoms on some important vegetable crops are given in Table 4:

Table 4: Specific Symptoms of N Deficiencies in Vegetable Crops

Crop	*Symptoms*
Beet root	Leaves pale green, turning dark reddish, purple later, habit of growth thin erect.
Broccoli, Brussels sprouts	Young leaves pale green, older leaves coloured orange, red to purple, followed by shedding.
Brinjal muskmelon	Stunted growth, leaves pale green, older leaves small & uniformally pale green, start getting bleached from margins inwards until finally entire leaf is bleached to pale white (general paling & stunting).
Cucumber	Growth stunted, fading of leaves to various shades of green & yellow. Stem slender, hard & fibrous. Fruits pale in colour, smaller& pointed at their blossom-ends.
Celery	Older leaves show yellowing followed by wilting and death.
Carrot	Leaves coloured light green, followed by yellowing; petioles weak
Cassava	First on the older leaves, lower leaves becoming chlorotic, plants appear stunted.
Lettuce	Leaves pale green, older leaves showing yellowing, followed by death and drying or firing

Contd...

Table 4–Contd...

Crop	*Symptoms*
Onion	The plants grow slowly, become stiff and upright in habit. Develop short, narrow leaves of light green; Later the tips of older leaves die, showing bleached yellowish colour. Seedlings grow slow, develop short leaves which are light green in colour. Tips turn yellowish later the plant becomes stunted.
Okra	Stunted growth, thin shoots, small and sticky leaves, abscission of leaves at advanced stages of deficiency. The plants ultimately turning yellow with hard & brittle foliage, premature bolting with small & tough pods.
Potato	Plant growth is restricted, plants turn pale to yellowish green in colour. Later chlorosis on margins of lower leaflets which loose chlorophyll & fade to pale yellow. Shedding results in reduced yields therefore poor quality tubers.
Radish	Retarded growth; leaves small, narrow thin yellowish. Roots small, imperfectly developed, faded reddish colour.
Sweet potato	Leaves yellowish green with reddish tinged areas, short petioles; defoliation of old leaves; tip leaves light green, small growth stunted.
Tomato	Growth is retarded, young leaves remain small & thin & the entire plant gradually becomes light green to pale yellow. Veins on the leaf turn yellowish green to finally purple especially on the under side. Stem becomes hard & fibrous acquiring the purplish discolouration as in veins. Flower buds discolour yellow & drop, fruit becomes small & stunted. Roots remain stunted, turn brown & die.
Cauliflower	Plants dwarfed, leaves small, young leaves pale, green, older leaves coloured purple followed by shedding, formation of head delayed.

Phosphorus

It is a fascinating plant nutrient and is involved in a wide range of plant processes- from permitting cell division, to the development of a good root system, and to ensure timely ripening of the crop. It is present in inorganic form as a component of ATP, RNA, DNA, certain enzymes and proteins and is involved in various energy transfer reactions and genetic information. It limits nodule development and nitrogen fixation, plant growth and seed development. Phosphorus is needed by young, fast growing tissues and performs a number of functions related to growth, development, photosynthesis and utilization of carbohydrates. It benefits crops in many ways, it gets a head start right from the seedling stage, produces deeper and proliferous roots which enable it to feed on a bigger soil volume for water and nutrients above all it produces more fruity cites. Phosphorus gives better resistance to drought spells as its longer roots can tap deeper soil layers for water when the surface soil is dry and increases plants tolerance to salinity. The net result of optimum phosphorus supply to the crop is a high yield of superior quality produce. It is an important structural component of a wide variety of biochemicals, including nucleic acid, coenzymes, nucleotides, phospho proteins, phospholipids, phytic acid, and sugar phosphates, with a major function in energy transfer. Phosphorus combines with organic compounds, forming relatively stable high-energy complexes such as adenosine diphosphate and adenosine triphosphate, which provide the mechanism for trapping, transporting and donating energy so that enzymatically catalyzed reactions can proceed (Hageman, 1969; Mengel and Kirkby, 1978).

An adequate supply of P early in the life of a plant is important in laying down the primordial for its reproductive parts. Large quantities of P are found in seed and fruit and it is considered essential for seed production. Phytin, compound of calcium and magnesium salts of phytic acid is the principal storage form of P in seeds. Phosphorus has long been associated with early maturity of crops. The quality of certain fruit and vegetables is said to be improved and disease resistance increased when these crops have a satisfactory P nutrition. Also the risk of winter damage with resultant poor yields can be substantially lowered by application of phosphorus.

Role in Plant

Phosphorus affects plants in many ways:

- ☆ It encourages plant cell division.
- ☆ Flowers and seeds do not form without it
- ☆ Phosphorus hastens maturity, thereby offsetting the quick growth caused by nitrogen.
- ☆ It encourages root growth and the development of strong root system.
- ☆ It makes potash (potassium) more easily available.
- ☆ It increases the plants resistance to diseases.
- ☆ It improves the quality of grains, root and fruit vegetable crops.
- ☆ Component of amino acids and chlorophyll.
- ☆ Enables meristematic growth.
- ☆ Essential for root growth during the early stages of plant life.

Deficiency

Phosphorus is non-mobile and non-leachable in soils but in growing plants is quite mobile. When deficiency start to develop in a plant, the absorbed phosphorus migrates from older to youngest parts, that is to the active meristematic regions, therefore, its symptoms appear first on older leaves like nitrogen. The beginning symptom is a loss of sheen or shine of older leaves and leaves appear dull. Gradually a bluish–green to reddish colour develops which can lead to bronze tints and red colouration. This "fall" colouration spreads to other portions of leaves as the deficiency progresses. Young leaves remain green during this time but small, often reduced to about one-tenth of normal size. Older leaves usually drop before necrosis or death occurs, but if they don't, necrosis begins at leaf tips and progresses towards the base (Dickey,1977). Seedling death and stunted growth are common. Its deficiency has a marked effect on retarding over all growth.

Its deficiency causes delayed starch production, low cambial activity and lack of succulence in plant tissues in vegetable crops. Phosphorus deficiency increases disturbance in metabolism which gives the wide toxic products within the plant (Wadleigh, 1949). Marginal necrosis may appear in leaves, petioles and fruit. In terms of general growth, the effects of phosphorus deficiency are very similar to those of nitrogen.

There is restricted root and top growth, retarded flowering and ripening and lowered quality in addition to yield decrease, are the general effects of phosphorus starved plants. In general, deficiencies of P in vegetable crops are less common and less devastating than N deficiencies (Table 5).

However insufficient phosphorus results in (1)purple colouring on the under surface, (2) reduces flower, fruit and seed production (3) susceptibility to cold injury

(4) susceptibility to plant diseases and (5) poor quality fruit and seeds. The visual deficiency symptoms on some important vegetable crops are given in Table 5.

Table 5: Deficiency Symptoms of Phosphorus in Vegetable Crops

Crop	*Symptoms*
Beet root	Leaves dull purple, small, with short petioles
Broad bean	Leaves are dark green; older leaves drop, stems are thin, blossoms are sparse
Brinjal	Dirty greyish green leaves and premature shedding, leaves small & greyish in the beginning, later turning to dirty greying green
Cassava	Uniform chlorosis developed by the lower leaves
Cabbage	Leaves are dull green with purple cast; especially on under side; they are small and firm but margins die
Carrot	Leaves are dull green with purple cast, older leaflets die, petioles are upright.
Lettuce	Leaves are dull green to reddish brown or purple; older leaves die, growth is stunted with poor head formation.
Onion	Stunted growth, leaves get mottled & tips of the older leaves wilt & die. The leaves soon become yellow and brown tissue as the necrosis advances towards the leaf base when dead, the leaves turn black.
Okra	Stunted growth, dark green foliage, no other characteristic
Potato	Retarded growth particularly early stages of development, plants are dull green in colour, stunted growth, foliage is crinkly and vegetative growth continues beyond the normal time of maturity. Petioles, leaflets & their margins turn upwards. Tubers with rusty brown lesions & isolated flecks in the flesh.
Pea	Leaves are bluish green and sparse; shoots are weak, thin and stunted

Contd...

Table 5–Contd...

Crop	*Symptoms*
Cauliflower	Stunted growth, leaves dull browning because of purple colour. Older leaves turning brown followed by tip drying, head forming late.
Radish	Leaves are reddish purple on under side; root development is poor; plant is stunted.
Tomato	Leaves small, veins on the underside leaves develop discoloured spots which later on turn purple, leaf tips purplish, stem slender & fibrous. foliage is sparse; plants are stunted. In acutely phosphorus- deficient plants purpling and stunting may be accompanied by small, sunken, circular necrotic areas, up to 3mm in diameter, on the leaflets. The spots sometimes assume a zonate character which may suggest those of early blight.

Deficiency Symptoms

- ☆ Retarded growth, underside of leaves & veins develop purple pigmentation.
- ☆ Wilting and death of the tips of the older leaves.
- ☆ Bronze to red/purple leaf colour and brown tissue as the necrosis advances towards the leaf base.
- ☆ Shoots are short and thin
- ☆ Older leaves die off rapidly
- ☆ Flower and seed production are inhibited
- ☆ Restricted root development
- ☆ Delayed maturity.

Potassium

Potassium is required for translocation of assimilates and is involved in the maintenance of water status of plant, especially the turgor pressure of cells, opening and closing of stomata and increase the availability of metabolic energy for the synthesis of starch and proteins. It is necessary for formation of sugars, starch, carbohydrates, protein synthesis and cell division in roots and other parts of the plant. It is essential for cell organization, hydration and cell permeability and also to maintain iron supply for chlorophyll synthesis, it is necessary for all root crops. Potassium controls enzyme systems that determine photosynthesis and respiration rates, carbohydrate metabolism, and translocation of organic acids and non protein N in plants. It is reported to be the primary activator for at least 46 individual enzymes (Evans and Sorger,1966).

K increases water holding capacity of plant tissues, succulence of vegetables, ship more easily, spoil less frequently and retain good condition for longer period. It helps to adjust water balance, improves stem rigidity and cold hardiness, enhances flavor and colour of fruit and vegetable crops. Besides, it increases peg formation, synthesis of sugar and starch and helps in pod growth and filling. In potash deficiency the plants are stunted and leaves are dark green. Young leaves have a rough surface and bronze colour. Its deficiencies may result in low yields, mottled, spotted or curled and scorched or burned look of leaves. Plants absorb more K than is needed. The greatest plant growth and highest starch and sugar content are most commensurate with high concentration of K in the soil solution or in the plant.

Potassium is usually, supplied from potassium chloride (KCL), potassium sulphate (K_2SO_4), or potassium nitrate (KNO_3). These sources of K are very soluble, even in the soil or growing medium and thus are leachable.

Role of Potassium in Plants

- ☆ It activates more than 60 enzymes and directly or indirectly involves in all major plant growth processes.
- ☆ Promotes photosynthesis, resulting information of carbohydrates, oils, facts and proteins.
- ☆ It is involved in movement of photosynthates to storage organs (seeds, tubers, roots, fruits).
- ☆ Improves efficacy of N fertilizers by enhancing production of proteins.
- ☆ Essential for formation of sugars in plants (potato and other tuber crops)
- ☆ It increases ability of plants to withstand stresses-draught, frost, pest, disease, lodging, poor drainage etc.
- ☆ Potassium regulates absorption of water by plant roots, helps development of healthy root system.
- ☆ It regulates respiration in plants.
- ☆ It improves quality of crops and prolongs shelf life of crop produce.
- ☆ Essential for efficient biological N fixation.
- ☆ Encourages a strong, healthy root system.
- ☆ Encourages the efficient use of carbon dioxide.
- ☆ Essential for the development of chlorophyll.
- ☆ Essential for tuber development.
- ☆ Improves the quality of seed, fruit and vegetables.
- ☆ Influences the uptake of other nutrients, *e.g.* Mg.
- ☆ Essential for flower and fruit formation.

Deficiency

Potassium is exceedingly mobile within plants since it is not complex with any organic substance and thus potassium deficiency is most clearly shown in older leaves. As with N and P, K- deficiency symptoms occur in older leaves first. The older leaves present pale chlorotic patches with the appearance of "burns" necrosis at the leaf tips and edges. The areas of dead tissue progress from the tip to the base and from the leaf margins towards the inter-veins area. The leaf tip tends to curve downwards. The root system is poorly developed, the internodes are slightly shorter and the stems are weaker, seed production is greatly diminished. Its deficiency symptoms frequently occur in vegetable crops late in the growth cycles, as it is translocated to developing storage organs. Yields and quality of vegetables are impaired. Marginal necrosis of older leaves is universal indication of potassium deficiency (Table 6).Often there is no chlorosis associated with K deficiency, but sections of leaves turn directly from living (green) to necrotic (brown). Necrosis usually begins at leaf tips or upper margins of leaves and progresses towards the leaf base, or it occurs as irregular spots throughout the leaves, being more severe at leaf bases; a combination of two patterns sometimes occur. Oily spots occur occasionally on the underside of basal leaves, which later develop into necrotic areas (Dickey,1977). In K deficient plants, toxic compounds such as diamines are reported. Potassium deficient plants, have smaller xylem vessels that tend to cluster in the center of the root, thus interfering with the flow of raw materials. When leaf K concentration fall below 1.5 per cent of the dry weight, deficiencies are likely to occur. Potassium has the most consistent effect on black-spot and tuber grown with low K levels are more susceptible to black-spot (Vander Zaag and Meijers,1969). Increased K levels often reduce specific gravity and it was supposed to effect in reducing black-spot was on tuber hydration (Van der Zaag and Meijers, 1970). An indirect effect of high K levels has been related to lower levels of tuber phenols and phenolase activity.

Deficiency Symptoms in General

Vegetables suffering from potassium deficiency show a reduced vigour, susceptibility to diseases & impairment of growth

- ✰ Earliest symptoms are as a slight yellowing of the oldest leaves followed by wilting & death of the leaf tips.
- ✰ Stunted growth
- ✰ Scorched look to the edges of older leaves (chlorosis) gradually progressing inward.
- ✰ Stalks are week and plants collapse easily.
- ✰ Shriveled seeds or fruits
- ✰ Brown spots sometimes develop on leaves.
- ✰ The tilted parts had a satiny texture & their entire leaf droop, remain somewhat inflated, gradually assuming a crepe-paper like appearance and then become about the same colour as the leaves that were deficient in nitrogen.
- ✰ The plants produce dark green tops and give low yields.

The visual deficiency symptoms on some important vegetable crops are given below:

Table 6: Deficiency Symptoms of Potassium in Vegetable Crops

Crop	*Symptoms*
Beans	Leaflets become chlorotic with necrotic brown areas at margins between veins; leaflets curl or drop downward
Beet root	Leaves bluish green around veins: surface is crinkled and margins curl downward; older leaves become yellowish to reddish brown between veins; and at margins; stems of older leaves show spots and strips; roots are dark and tend to rot; are poorly developed.
Cabbage	Bronzing starts from the leaf edges, progresses towards the center. Brown spots develop on lamina & the edges start curling & drying. Leaves show yellowing; heads are soft, puffy and small.
Cauliflower	Leaves dull- bluish, marginal scorching of leaves followed by yellowing and interveinal chlorosis, head size much reduced
Carrot	Leaves are slightly chlorotic followed by browning; roots are spindly; growth is short.
Celery	Leaves become dark green, curling of leaf lets and brown colouration caused by necrosis; leaf stems are short, with necrotic areas.
Cow pea	Leaf lets are more or less mottled; followed by necrosis and ragged appearance.
Cucumber	Leaf becomes bluish green near veins: leaf margins show bronzing and necrosis, young leaves are puckered or crinkled; fruit is constricted at stem end; growth is slow.
Cassava	Reduced plant growth, in severe cases-purple, molting of older leaves, curling up of leaf margins & chlorosis & necrosis of leaf & margins are seen (Asher,1975).
Onion	Older leaves first show slight yellowing, followed by wilting and death, appearing like crepe paper; dying and drying start at tips of older leaves; The plants produces dark green tops & bulb formation is poor. The tilted parts had a satiny texture, & their entire leaf drooped, remain somewhat inflated.

Contd...

Table 6–Contd...

Crop	*Symptoms*
Pea	Growth stunted, edges of lower leaves become brown & the seed becomes thick having tough seed coat..
Potato	Plant growth retards, shortens the internodes, leaves are first bluish green, older leaves become yellowish followed by necrosis and browning starting from tips and margins leaf lets are cupped and crowded together; stalks are slender with short internodes and may collapse prematurely; tuber flesh is bluish. Leads to hollow heart
Radish	Leaves are reddish purple on underside; root development is poor; growth stunted; shortens the internodes.
Tomato	Slow growth, plants develop dark bluish-green colour. Young leaves finely crinkled, older leaves turn dark bluish green at first later yellowish-green tint along the margins. Discolouration towards the center of the leaflets causing bronzing of the tissue. Stem becomes hard, fail to increase in thickness & remains thin & brown. Fruits are soft, blotchy, ripen evenly. Tomato fruits are low in total solids, sugar, acid, carotene & lycopene contents. The disease is known as blotchy ripening.
Okra	Suppression of leaf formation, leaf margins turning brown yellow followed by brown colouration, scorching of leaves, abscission of scorched leaves.
Brinjal	Retard growth is lesser in comparison with N.& P. leaves normal green but smaller in size with crinkled surface. Small whitish nerotic spots on entire lamina in older leaves. Interveinal chlorosis of young leaves followed by yellowing & premature shedding.
Lettuce	Chlorosis, followed by random necrotic lesions on the leaves
Carrot	Roots are less sweet and flesh does not have the desired luster and lack quality

Factors Affecting Availability of Micronutrients

In vegetables production, micronutrients play a catalytic role in nutrient absorption and balancing other nutrients. Micronutrients also play important role in enzymatic activities and synthesis. Their deficiencies result in huge reduction in yield and quality. Boron (Bo), Molybdenum (Mo). Zinc (Zn),Cupper (Cu), Iron (Fe) and Mangnese (Mn) may be applied as foliar spray to correct the deficiency, improve the production and quality of vegetables. Most micronutrients are required in a trace amounts, their deficiencies causes poor plant growth, restrict yield and in excess toxic to plants. Hence, micronutrient application should be done carefully for desirable result after ascertaining their deficiency symptom/plant tissue analysis.

The most important factors which affect the availability of micro-nutrients in soil are pH, organic matter, $CaCO_3$, clay minerals and nutrient interaction etc.

The availability of all micronutrients except Mo, increases with a decrease in pH. The presence of organic matter may promote the availability of certain elements notably Mn, B, Mo, Zn etc. where as it fixes and reduces the availability of Cu. Calcium carbonate decreases the availability B, Zn, Fe, and Mn and increases the availability of Mo. In calcareous soil, all these nutrients are less available. In addition to the direct effect of $CaCO_3$, the lime induces deficiencies of Fe, Zn, Mn etc. are due to the effect of bicarbonates. Lime induced Fe and Zn deficiencies are very common. The heavy metals affect the uptake and translocation of Fe. Phosphorus induces Zn deficiency in high Phosphorus soil and due to high P application excess Mn application can result in Fe deficiency.

Some examples showing effects of micro-nutrient deficiency on vegetables and the improvement in their quality through the application of micro-nutrients are given below.

Quality or Nutritive Value	*Micronutrient*
Decrease in vitamin A content of spinach	Manganese deficiency
Increase in vitamin C content of tomatoes	By application of manganese to soil
Decrease in vitamin C of tomato and vegetables	Molybdenum deficiency
Increase in vitamin C in Chinese lettuce, Chinese red peppers and tomatoes	By application of zinc, magnesium and nickel
Increase in vitamin B and biotin content of beans	By application of small quantities of selenium
Increase in thiamine and niacin content of tomatocs and turnips	Application of boron and molybdenum
Decrease in arginine of the protein in cauliflower	Decrease in molybdenum

Secondary Nutrients

Calcium, magnesium and sulphur are known as the secondary nutrients and their requirements by crop is intermediate between primary and micro-nutrients.

Calcium

Calcium (Ca), is constituent of middle lamella of cell walls as Ca- pectate, plays a role in cell wall rigidity by affecting IAA levels, stimulates the effect of ADP as an energy acceptor, thus, affecting respiration and ameliorates toxicities of other ions and organic acids (Evans and Sorger,1966). It ensures membrane permeability, pollen germination, activates the number of enzymes for cell division and takes part in protein synthesis and carbohydrate transfer. In its physiological effects, calcium is usually regarded as counter part to potassium. It influences the stability of structure of protoplast and owing to its dehydration properties, it opposes the plasma-expanding action of potassium. The calcium requirement is very high especially for gynophores development and pod filling. Since it plays a major role in cell division, its deficiency appears in meristematic and growing region of plants. Moreover, once it is absorbed, calcium is not mobilized from the older leaves, hence its deficiency occurs on the fresh and emerging leaves.The calcium ion (Ca^{2+}) is transported exclusively in xylem tissue upwards with transpiration stream but its downward movement from leaves through phloem is practically nil (Mengel and Kirkby, 1978).

It is generally encountered in acid soils. In cassava, calcium deficiency leads to tip burn and deformation of the upper leaves. Normal calcium levels of upper fully expanded leaves range from 0.6 to 1.5 per cent (Asher *et al.*, 1980). Liming is the common practice of neutralizing soil acidity and there by increasing the efficacy of native and applied nutrients. It also supplies Ca to the plants. Calcium sources are best broadcasted and incorporated into the soil prior to planting. The use of single superphosphate which contains 20 per cent Ca may contribute to the low incidence of calcium deficiency under field conditions. Excessive use of lime may lead to the induction of deficiencies of potassium, magnesium, iron, zinc and copper.

Calcium is immobile in the soil but is not retranslocatable within plants, and deficiency is shown in young leaves, shoots and the growth zone of the root. Deficiency may occur in acidic soils and highly leached sandy soils. The deficiency symptoms usually occur in younger portions of the plant first. Terminal leaves become small without a patterned chlorosis and older leaves become thick and brittle. Stem tips die and further growth ceases. The young leaves on the shoots, especially those around the apical bud, are malformed, showing curling and curl around the growing tips and the apical meristem eventually dies. If the axillary bud develop, they also die in the end. In the already developed young leaves large chlorotic patches appear at the edges. The roots turn mushy and die. Excessive levels of Na, K or Mg in irrigation water, fertilizers or dolmite may hinder the uptake of Ca leading to Ca deficiencies. A number of Ca deficiency disorders of vegetable crops have been reported (Table 7). Black heart of celery, brown heart of endive, tip burn of lettuce and internal tip brown of cabbage are growing point disorders due to calcium deficiencies. In Brussels sprouts, calcium deficiencies is found to occur in the axillary buds and not on the growing terminal point. Blossom-end rot of tomato,capsicum and watermelon appears first as water soaked lesion, a series of lesions at the blossom-end of the fruit during fruit enlargement and maturation (Maynard,1979). Where in case of carrot and parsnip, calcium deficiency causes a cavity in the root phloem immediately below epidermis. These disorders can be controlled to great extent by promoting cation balance and use of foliar sprays of soluble Ca salts. Calcium in addition to reducing the severity of physiological disorders, alters intercellular and extra-cellular processes which intimately associated with senescence and fruit quality. It confers rigidity to cell walls and serve as intermolecular binding agent that stabilizes pectin-protein complexes of the middle lamella. Calcium, with concentrations ranging from 0.2 to 1 per cent in plant tissues, is also essential to plant life. Calcium deficiency manifests in the failure of terminal buds and apical tips of roots to develop. Also lack of calcium results in general breakdown of membrane structures, with resultant loss in retention of cellular diffusible compounds. Disorders in storage tissues of fruits and vegetables frequently indicate calcium deficiency (Tisdale *et al.*, 1985).

Role of Calcium in Plants

- Calcium is essential for the development of growth tissue (tips).
- It maintains the cell integrity & membrane permeability.
- It enhances pollen germination.
- It activates the number of enzymes for cell division and takes part in protein synthesis and carbohydrate transfer.
- Constituent of cell walls.
- Essential for cell division, especially in roots.
- Acts as a detoxifying agent by neutralizing organic acids in plants.
- Prevents bitter pit in apples and cork spot in pears
- It improves storage life.
- Prevents black heart of celery, brown heart of endive, tip burn of lettuce and internal tip brown of cabbage.
- It influences the stability of structure of protoplast and owing to its dehydration properties, it opposes the plasma-expanding action of K.

☆ The Ca requirement is very high especially for gynophore's development and pod development.

The visual deficiency symptoms on some important vegetable crops are given in Table 7.

Table 7: Deficiency Symptoms of Calcium in Vegetable Crops

Crop	*Symptoms*
Beans	Blackening and death of the plant
Beet root	Leaves are pale green around margins, curled towards upper surface, necrotic and ragged. Rots are forked and turned.
Brinjal	Light green colour of young leaves along with necrotic spotting.
Cabbage	Leaves rolled up at margins which are ragged and discoloured; white in narrow band, followed by necrosis at times; death of growing point.
Cauliflower	Leaves rolled up at margins which are ragged and discoloured, white in narrow band, followed by necrosis at rins.
Carrot	Leaves show some chlorosis, necrosis, or scorching and finally die, foliage is sparse.
Cassava	Tip burn and deformation of upper leaves.
Pea	Youngest leaves are curled and tough; lower leaves are chlorotic. Plants are short and die prematurely. Root tips die.
Potato	Young leaves are small and pale green, rolled towards upper surface, with marginal necrosis; bud dies; tubers are dwarfed and useless, with dead spots in pith region. Light green bands along the margins of the young leaves & the terminal buds often have a wrinkled appearance. Severe cases young leaves remain folded & later die. Axial growth ceases; medullary region in tuber shows dead spots of diffuse brown discolouration.

Contd...

Table 7–Contd...

Crop	*Symptoms*
Radish	Leaves of young plants show narrow white bands at margins; there may be some interveinal chlorosis. Also marginal wilting, necrosis and rolling up.
Sweet potato	Young leaves are light green, some older leaves may show reddish areas and necrosis.
Tomato	Young leaves of terminal growth turn yellow, brown or purple and become necrotic. Apparently calcium is not transferred readily from lower to upper leaves. Terminal flowers die. Plants lack turgor and are weak and flabby. The continued deficiency results in flabbiness, death of the terminal bud, short, stubby and brown, much branched roots, and blossom-ends rot.

Deficiency Symptoms

Calcium deficient plants develop yellow leaves which are weak, flabby, lack firmness and terminal buds die. The roots remain short, highly branched, stubby, bulbous and dark brown in colour (Table 7).

- ☆ Terminal leaves become small without a patterned chlorosis and older leaves become thick and brittle.
- ☆ Chlorosis of the young foliage and white colouration to edges.
- ☆ Stem tips die and further growth ceases.
- ☆ Growing point may shrivel up and die.
- ☆ Grey mould (botrytis) infection to growing point
- ☆ Distorted leaves with the tip hooked back.
- ☆ Death of root tips and damaged root system (appearing rotted).
- ☆ Buds and blossoms shed prematurely.
- ☆ Stem structure weakened.

There is some relationship between fruit calcium and fruit quality in terms of texture and storage potential (Brady, 1993). As Marschner (1995) stated that low tissue contents of calcium in freshly fruits increase losses caused by enhanced senescence of the tissue. Even a small increase in calcium content of the fruits can be effective in preventing various storage disorders. Calcium bound as pectate in the middle lamella is essential for strengthening cell walls and plant issues. Application of calcium as a pre-harvest treatment can either reduce internal browning or totally eliminate the occurance of this disorder.

Magnesium

Magnesium (Mg), the metallic portion of the chlorophyll molecule is required by many glycolytic enzymes and is an activator of enzymes containing sulfhydryl group, especially those involving P metabolism. It is a constituent of chlorophyll and chromosomes. It acts as a bridge between pyrophosphate structures of ATP or ADP and the enzymes molecules and it serves as a cofactor in most of the enzymes activating phosphorylation. It is also a constituent of polyribosomes which are essential in protein synthesis. Magnesium deficiency is usually found in acid soils and is induced by high content of potassium in the soil. The striking symptoms of magnesium is an internal chlorolysis of lower leaves. Continuous cultivation of cassava in lateriate soil with a high potash application (100 kgK_2O/ha) has resulted in magnesium deficiency in cassava variety (H-1687) Sreevisakham. In case of potato due to magnesium deficiency, lower leaves show chlorosis on the margins and at the tips. This chlorosis increases and the entire interveinal area may turn yellow. The deficiency of magnesium is also a problem of sandy and strongly acid soils as under high rain fall, the Mg is leached out more easily in these soils. It is taken up by plant as the divalent cation (Mg^{2+}) from the soil. A magnesium ion resembles calcium in its behaviour on ion exchange and as an exchangeable ion. Mg is second in abundance to calcium, but with increase of pH in alkaline soil, Mg becomes non-exchangeable. It influences the phytohormone balance and nitrate reduction. Magnesium is important in chlorophyll formation and is involved in photosynthesis.

Role in Plant

- Essential constituent of chlorophyll and chromosomes.
- It is also a constituent of poly-ribosomes which are essential in protein synthesis.
- It influences phosphorus mobility
- It influences potassium uptake by roots.
- Helps the movement of sugars within the plant.
- An activator of enzymes containing sulfhydryl group.
- It plays a vital role as a phosphate carrier.

Deficiency

Its deficiency is among the most prevalent nutritional problem in vegetable cultivation, because Mg is not commonly provided in most fertilizer programs, is readily leached with the watering regimes, and occurs low in sands and organic materials used as growing media. It is frequently deficient in sandy, well drained soils. Magnesium is soluble in the soil solution but held to the clays and organic matter.

Magnesium being an important constituent of chlorophyll molecule, therefore, its deficiency symptoms are fairly specific in most pinnately (netted) veined leaves. It deficiency causes chlorosis due to the failure of the synthesis of chlorolysis. (Table 8). Magnesium is mobile, so its deficiency appear first on the older leaves as interveinal chlorolysis and redistribute to younger leaves. Chlorosis which begins at upper or distal margins of lower leaves, first progressing inward and downward in an arc like pattern. The chlorosis is actually a bronze yellow colour, which gave Mg deficiency the original name of bronzing disease. The chlorotic pattern progresses leaving a V – shaped tip of green on the leaf tip and an inverted V- shaped green area at the leaf base. As the deficiency becomes more severe, the tip loses its green colour first, followed by the basal section, and by this time necrosis begins in the upper margins where chlorosis was first apparent. The leaf tip and margins may curl up. The leaves do not dry up. Necrotic areas spread in the same manner as the chlorosis and are usually rusty brown in appearance (Dickey and Joiner, 1966).When the leaf concentration of Mg falls below 0.2 per cent of dry weight, deficiency symptoms appear. Its deficiency causes vascular browning and it was corrected by using 1.5 per cent Mg SO_4 as foliar spray (Winsor *et al.*, 1965).

Massey and Adams (personal communication) encountered magnesium deficiency symptoms in tomato when a change was made from well water containing 7 ppm magnesium to municipal water containing 1.5 ppm. Increasing the concentration of magnesium fertilizer overcame the disorder. Alternatively, a foliar application of magnesium salt (*e.g.*, 1 per cent $MgSO_4.7H_2O$) will alleviate symptoms. At the time of root death, when there is a heavy fruit load on the plant, young leaves often show symptoms of magnesium and iron deficiency (interveinal chlorosis). It is possible that the roots do not absorb divalent cations at that stage.

The visual deficiency symptoms of some important vegetable crops are given in Table 8:

Table 8: Deficiency Symptoms of Magnesium in Vegetable Crops

Crop	*Symptoms*
Beet	Older leaves show chlorosis and reddish tinting between veins.
Brinjal	Inverted "V" shaped interveinal chlorosis, more marked in central areas between veins. The fruits are small and may shed.
Cauliflower	Chlorosis of veins leading to the mottling of older leaves, chlorotic areas may form perforation.
Cabbage	Chlorosis, mottling & puckering of lower leaves. 'Marbling' occurs, in severe cases mottling is more pronounced which develops white, bronzed or pale yellow areas around the edges and center. Later these areas decay and drop out.
Carrot	Leaves are light coloured; tips or lobes show light yellow or brown spots, old leaves are severely chlorotic.
Okra	Interveinal yellow spots appear on leaves; these increase in number until only the veins remain green.
Onion	Leaves die back at tips; foliage dies prematurely, growth is slow. Onions develop irregular, elliptical white areas near the ends of the leaves, which later disappear in a general breakdown of the affected tissue.
Pea	Leaf tips are browned, leaves die prematurely.

Contd...

Table 8–Contd...

Crop	Symptoms
Chillies	Leaves turn pale green followed by interveinal yellowing; lower leaves may drop. Plants are small. Fruits sparse and undersized
Radish	Older leaves become chlorotic between veins.
Tomato	Older leaves become brittle & the intervenial tissue becomes chlorotic.Symptoms appear first on the lower leaves & the condition progresses upward. Yellowish discolouration between the veins which become intense with increasing distance, such leaves turn brown and bronze. Petioles become etched, and tend to hang down, the stalk, leaf margins turn up. Stalks are slender. Roots are long, with few branches & causes vascular browning.
Cassava	Produces interveinal chlorosis of lower leaves.
Celery	Chlorolysis
Turnip	Leaves develop brown improperly formed areas around their margins which dry up and breakdown while inner tissues become chlorotic and mottled.
Cucumber, squash, muskmelon	Typical mottling and browning of leaves occur.
Potato	Foliage pale in colour, lower leaves first affected, which show a loss of green colour at the tips and margins, chlorosis progresses towards the center of the leaflets. Severe cases middle portions of leaflets between veins become chlorotic and finally filled with small brown dead areas. Leaflets are badly affected, show bulging between the veins & become thick & brittle. Finally chlorotic leaves turn brown, die & often drop off.

Deficiency affects the earliness, uniform maturity, size of roots and fruits, and other market related qualities.

- ☆ Leaves develop yellow margins and pale green\ yellow blotches between veins (interveinal chlorosis).

- Blotches become yellower and eventually turn brown.
- In final stages leaves are small and brittle, edges turn upwards.
- In vegetables, plants are coloured with a marbling of yellow with fints of orange, red and purple.
- Stems are weak and prone to fungal attack.
- Premature leaf drop.
- There is interveinal chlorosis of lower leaves.

Sulphur

Sulphur (S) plays a major role in the synthesis of the amino acid cysteine, which in turn is vital in the formation of cystine and methionine, activates certain proteolytic enzymes, such as papainases (Wilson,1962), and is a constituent of certain vitamins and coenzymes, such as lipoic acid and coenzyme. It is essential component of vitamins like thiamine (B_1) and biotin. Sulphur like nitrogen is involved in low- energy bonding and protein synthesis. It forms thiol bonds analogous, energetically, to the N peptide bonds, sulfhydryl group (SH) are thought to be important in the hardening of protoplasm to cold and drought. In energy transfer sulphur can function in a manner similar to phosphorus.

It improves nodulation and pod yield, reduces incidences of diseases and is important as Phosphorus. It increases chlorophyll and decreases chlorosis in calcareous soil by increasing availabilities of micronutrient in soil (Singh *et al.*, 1990).It is essential to plant nutrition. In general plants contain as much S as P, the usual range being from 0.2 to

0.5 per cent on dry weight basis. Sulphur ranks in importance with basis. Sulphur, ranks in importance with N as a constituent of the amino acids, cysteine, cystine and methionine in proteins that account for 90 per cent of S in plants.

It helps in the formation of chlorophyll, the green substance in the leaves that permits photosynthesis. Plants produce starch, sugars, oils, fats, vitamins and other vital compounds through photosynthesis.

It helps in protein production, primarily because S is a constituent of three essential amino acids *viz.* cysteine, cystine and methionine, which are the building blocks of protein. About 90 per cent of plant S is present in these amino acids.

It is involved in the formation of glucosides and glucosinolates (mustard oils) and sulphydral (SH-) linkages which are the source of pungency in onion, oil etc.

It activates enzymes, which aid in biochemical reactions within the plant. Sulphur is known to be a principal constituent of allyl propyl disulfide- responsible for bulb pungency.

It increases the crop yield and improves the produce quality, both of which determine the market price. In plants suffering from sulphur deficiency the rate of plant growth is reduced. Generally the growth of the shoot is more affected than root growth. Frequently the plants are rigid and brittle and the stems remain thin. In cruciferae the lateral extension of the leaf lamina is restricted and the leaves are rather narrow.

In peas the young leaves turn pale followed by chlorosis

of interveinal areas first on young leaves and then in middle and old leaves. Root nodulation, flowering and yield is reduced. In potato there is pronounced inward curling of youngest leaves along with considerable yellowing of the stems and general yellowing of the plant is observed. While in French beans plants have short internodes, few and smaller leaves. The entire foliage appears pale green. Eventually plants have poor growth and yield. Plants are smaller and light green in colour than normal in case of tomato. Some times yellowing may occur in various plant parts. While in severe cases deficiency, petioles and stems show a marked reddening.

Members of cruciferae family may contain over 1 per cent sulphur and legumes are also relatively high in sulphur. Oils from some plants, particularly those of cruciferae and onion are rich in sulphur.

Role in Plant

- It is component of amino acids, proteins and oils.
- It is an essential components of vitamins and coenzymes such as Lipoic acid & coenzyme A.
- Aids the stabilization of protein structure.
- It activates certain proteolytic enzymes such as papainases which aid in biochemical reactions within the plant.
- It is also a part of certain vitamins such as biotin and thiamine (B_1).
- It increases crop yield and improves produce quality.
- It is essential for growth and development of all crops.

☆ Formation of chlorophyll, the green substance in leaves that permits photosynthesis.

Deficiency

It is highly immobile element, so that sulphur deficiency unlike phosphorus and nitrogen deficiency is located in the younger shoots and parts. The young leaves show a general chlorosis, both along the veins and between the veins and the younger the leaf, the yellower it is. The leaves grow less than in control plants and the entire shoot is pale. In some plants (straw berries), anthocyanin colourings may be present. There is newer necrosis in these young leaves.

Being soluble in most of its forms and is therefore, leachable from the soil. It is immobile in plants, and deficiency symptoms occur in the young leaves first and are identical to those for N deficiency (*i.e.*, a general yellowing of the entire leaf) but occur on younger rather than older leaves, which is just opposite of N deficiency (Lunt *et al.*, 1964). Sulphur deficient leaves are pale green to yellow in colour and the deficiency symptoms are associated with concentrations less than 0.3 per cent and in young fully expanded leaf blades of 12 weeks old plants (Asher *et al.*, 1980). Sulphur deficiency can be controlled by application of 10- 20kg sulphur per hectare as element sulphur or gypsum. Application of 50 kg sulphur per hectare with sulphur containing fertilizers (SSP, Ammonium Sulphur and ammophos). Crop raised on S-deficient soils are small and spindly with short and slender stalks. Their growth is slow, yield is low, quality is inferior and nitrogen fixation in legumes is reduced.

Sulphur deficiency like nitrogen is expressed as stunting

and general plant yellowing, stems are thin. Although sulphur is mobile in plant, redistribution from older to younger leaves is not as pronounced with sulphur as with nitrogen and firing of lower leaves does not commonly occur. The lower leaves of plant become yellowish green. The stems are elongated, hard, woody, roots well developed and extensive. In general plants suffering from S deficiency the rate of plant growth is reduced. Frequently the plants are rigid and brittle and the stems remain thin.

Deficiency Symptoms

In sulphur deficient plants, the lower leaves become yellowish green, the stems are elongated, hard and woody and roots well developed and extensive. The deficiency symptoms of some vegetable crops are given in Table 9.

Table 9: Deficiency Symptoms of Sulphur in Some Vegetable Crops

Crop	*Symptoms*
French bean	Plants have short internodes, branching is poor, plants have bushy appearance. Flowers are drastically reduced & pods have shrunken seeds.
Pea	Young leaves turn pale followed by chlorosis of interveinal areas first on young leaves & then in middle and old leaves. Root nodulation, flowering and yield is reduced
Potato	Pronounced inward curling of younger leaves along with considerable yellowing of the stem and general yellowing of the plant is observed. Stem woody and root development may decrease.
Tomato	Plants are smaller & lighter green in colour than normal. Yellowing may occur in various plant parts. In case of severe deficiency, petioles and stem show a marked reddening, root development may decrease. The stems tend to elongate abnormally and remain below normal in diameter. There is a tendency to increase in carbohydrate content.

- There is uniform yellowing of new and young foliage.
- Plants deficient in S are small and spindly with short and slender stalks.
- Retards growth and delays maturity for example in pea.
- The growth of the shoot is more affected than root growth.

Boron

It plays major role in flowering and fruiting processes, pollen germination, cell division, nitrogen metabolism, sugar translocation, carbohydrate metabolism, active salt absorption, hormone movement and action on metabolism, pectic substances, maintenance of conducting systems, water relations, fat metabolism, buffer action, precipitation of excess cations and regulation of older elements, and that is a membrane constituent. It facilitates translocation of sugar and fat synthesis and is important for RNA (uracil) synthesis, cell division, differentation, and maturation. The boron is transported primarily in xylem and is relatively immobile in neutral to alkaline and highly weathered soils. The factors influencing boron deficiency are soil low in boron, organics in soils, low humic gley, moderate to heavy rainfall, dry weather and light intensity. In beans its deficiency symptoms consists of chlorosis and stunting and thus may be confused with manganese deficiency. Variety differ in the degree to which they exhibit symptoms on copper-deficient soils. The deficiency may be corrected by disking in about 125.0 kg/ha of copper sulphate/ha or by spraying the crop with copper fungicide. Boron causes internal break down, canker or dry rot of beet. It decreases the expression of potato wart disease (*Synchytrium endobioticum*) and club root (*Plasmodiophora brassicae*).

Role in Plant

- ✰ It is neither an activator nor a constituent of an enzyme complex.
- ✰ Its deficiency leads to abnormal differentiation of young tissues and accumulation of soluble carbohydrates and nitrogen at the expense of protein nitrogen.
- ✰ Its absence prevents pollen germination in many plants.
- ✰ It is believed to influence cell development by control of sugar transport and polysaccharide formation.
- ✰ Another function attributed to this element is combining with the active site of phosphorylation to inhibit starch formation.

Deficiency

Boron is not mobile, so deficiency appears in the young leaves. Where as deficiency is characterized first by internodes becoming noticeably shorter, with thickened stems that become tough and brittle and leaves that become small, stiff and ridged. Black, sunken necrotic spots develop in stems just below nodes. The leaves surrounding the terminal bud turn light green at the base, and eventually fall off. Later growth shows distorted leaves that are stunted and fragile. Eventually the growing tips suffer necrosis and the apical meristem dies along with the tip of the shoot. Fleshy organs may not internally show necrosis. Black, sunken necrotic spots develop in stems just below nodes. In vine crops, nodal roots become thickened, stubby and slough off. Vines develop a characteristic curling at the nodes producing a "pig tail" appearance. Boron deficiency destroys genotropic responses, cause plants to grow laterally and in abnormal way. Terminal leaves of a affected plants are very small, puckered, thickened, and blunt and have irregular steaks of chlorotic areas interveinally (Dickey,1977). It causes hollow stems of cruciferous crops like cabbage, cauliflower and broccoli, due to inadequate supply of boron. Brown heart of turnip appears first in irregular shaped water soaked areas. Black heart or canker of beet occurs in internal or external necrotic areas in the root. In celery, cracked stem appears as cross splitting of vascular bundles in the fleshy petiole resulting from a decrease in collenchyma cell wall thickness and in the number of cellulose lamella (Lewitt and Smith,1975).

In tomato, disorder appears as open locules in the fruit, internal browning, stem resetting when boron levels in the fruit are about 6ppm and when the leaf boron is about 15ppm on dry weight basis. Boron deficiency occurs in most vegetables if its levels is less than 30 ppm on dry weight basis. Its deficiency causes brown heart, heart rot or hollow center in turnip, radish and beet, where as in cabbage the deficiency symptoms (Table 10) first appears on fifth to tenth leaves from the outside of the plant when head begins to form and the inner parts of the midribs become cracked and rocky. Boron causes internal break down, canker or dry rot of beet. This trouble is serious in seasons and on soils more alkaline pH than 7.0: Borax at the rate of 62.5 kg per hectare is effective in controlling the disorder. In tomato there is cracking of fruits. It affects the formation and the utilization of carbohydrates, causes pitted and corky areas in fruits. The fruits formed are malformed and deformed in shape, and there is uneven ripening. In lettuce deficiency may develop tip-burn symptoms, which suggests that the addition of boron may contribute to the prevention of tip burn (Crisp *et al.,* 1976). In fenugreek boron deficiency causes flowering and reproductive failure, resetting of terminal buds, small leaves and chlorolysis. The two most important disorders namely cracked stem of celery, heart rot of sugar beet & turnip, leaf roll of potato and brown curd of cauliflower occur due to its deficiency and result in complete crop loss. Brown heart caused by its deficiency is quite common in table beets, turnip, radish and other root crops. All indicating cell wall integrity problems have been traced to boron deficiency.

The deficiency symptoms of some important vegetable crops are given in Table 10.

Table 10: Deficiency Symptoms of Boron in Vegetable Crops

Crop	*Symptoms*
Table beet, turnip & radish	Plant stunted, leaves remain small, less in no. and variegated due to development of yellow & purplish red blotches; roots develop dark spots in the thickened part, become distorted. Remain small, unhealthy, greyish with wrinkled and cracked surface; roots show brown heart, develops large water soaked area or even a hollow heart, necrosis of short apex and heart rot of sugar beet.
Tomato	Reduced root growth, swollen hypocotyls & cotyledons, Young leaves at seedling stage turn purple, growing point of stem becomes black; Petioles & midribs become extremely brittle. The stem becomes stunted & the terminal shoot curls inward, turns yellow & dies. Roots

Contd...

Table 10–Contd...

Crop	Symptoms
	how poor growth and turn yellow or brown; fruits develop dark or dried areas. Seedling and the true leaves turn to a distinctly purple colour. The stem becomes stunted and the terminal shoot curbs inwards, yellows and dies.
Lettuce	Malformation of growing leaves, spotting & burning of leaf tips and death of growing points. Marginal growth ceases & results in folding of leaflets. Increasing no. & size of leaf spots. Older leaves are least affected. Plants deficient in boron develop tip burn symptoms.
Cauliflower	Small concentric water soaked areas on stem and in the center of the small branches. In the center browning heart is conspicuous. Both inner & outer portions of the head are affected which develop a bitter flavor. Smaller leaves of affected plants deformed and stunted. Curds remain small in size & develop bitterness. The leaves get elongated, brittle & chlorotic at margins, which turn downwards. The stem becomes hollow and brittle.
Cabbage	Water soaked areas appear at the base of the head on the stem, which usually die out The symptoms are dwarfing and distortion of leaves with concentric water-socked areas, surface browning, bad flavour and interveinal breakdown in pith.
Onion	Plant stunted, giving a distorted look; leaf colour varies from dark grey green to blue green, young leaves develop conspicuous yellow and green mottling. Shrunken areas appear, in ladder like transverse cracks on the upper sides of basal leaves. Leaves become stiff & brittle. It is needed in minute quantities i.e 6.25 to 25kg per hectare.
Sweet potato	Restriction of terminal growth of vines & shortening of internodes. Petioles get curved & terminals become stunted & stunted. Older leaves turn yellow and premature shedding occurs. Tubers are misshapen, dumb-bell shaped, their skin becomes rough & leathery. Affected surface shows cankers, sometimes covered with a dark exudates. Necrosis in tuber flesh. Internal areas are necrotic throughout the fleshy part of the tuber but more so in the cambial zone near the periphery.

Contd...

Table 10–Contd...

Crop	Symptoms
Potato	Growing point dies; tips of terminal shoots are stimulated into a characteristic growth of lateral buds. Internodes remain short & give plant a bushy appearance. Leaves become thick & margins roll upwards. Roots short, thick and have a brown appearance, root tips die. Tubers small & have ruptured surface.
Celery	Cracked stem with lesion on the inner & outer surface of the petioles & over the vascular bundles. The adjoining epidermis curl outwards followed by a dark brown colour of the exposed tissue
Turnip	The interior of the affected roots becomes water soaked and may develop brown spots and cracks. The leaves of affected plants are dwarfed and curled. Cracked and hollow roots and brown heart.
Brinjal	Death of apical bud coupled with browning of roots.
Radish	Leaves turn light bluish-green and the petioles are brittle and curled down; roots remain light in colour, round, elongated and severely cracked. Apical buds and phloem develop necrosis.
Pea	Short, thick roots with enlarged apices and suppressed growth of secondary roots are the symptoms in this plant
Broad bean	Shoot apex withers and internal tissue disorganizes, particularly in roots
Fenugreek	Flowering and reproductive failure, resetting of terminal buds, small leaves & chlorosis

Molybdenum

Molybdenum (Mo) is a part of the nitrate reductase enzyme, which is involved in the initial step of nitrate reduction to nitrite (Mengel and Kirkby,1978). It acts as a metal component of enzymes nitrogenase and nitrate reductase that are closely related to nitrogen metabolism.

The only known use of Mo is the enzyme nitrite reductase and nitrate reductase, where it acts as electron carrier between oxidized and reduced states. Of all the micro-nutrients, Mo is needed in the least amount. Its deficiency disrupts the nitrogen metabolism and the plant shows nitrogen deficiency because of role of Mo in nitrogen fixation. The factors influencing its deficiency are low soil Molybdenum, acid soil, high organics and high free iron. Though Mo availability increases with pH, the deficiencies are quite likely to occur on soils with high pH. Also. In calcareous soil, the Mo deficiency causes 13-19 per cent yield losses. Antagonistic interaction of copper with molybdenum and acid nature of soil aggravate the deficiency of molybdenum.

The nitrogenase and nitrate reductase are the main enzymes influenced by molybdenum of all the micro-nutrients, though Mo is needed in the least amount, the Mo deficiency disrupts the nitrogen metabolism and the plant shows nitrogen deficiency because of the role of molybdenum in nitrogen fixation. The factors influencing molybdenum deficiency are low soil molybdenum, acid soil, high organics and high free Iron. Though Mo availability increases with pH, the deficiencies are quite likely to occur on soil with high pH also. In calcareous soil, the Mo deficiency causes 13- 19 per cent yield loss.

Role in Plant

- Essential for nitrogen fixation in the soil by Azotobacter and by the root-nodule bacteria.
- Involved in nitrate reduction in micro-organisms and higher plants.
- Its deficiency leads to loss of ascorbic acid.

Deficiency

Molybdenum deficiency in most broad-leaved plants is called "strap leaf," since affected leaves are reduced in width much more than in length. Mo is not mobile within the plants: therefore, symptoms, occur on young leaves first. The affected leaves are rough, thick, and leathery with margins irregularly wrinkled or buckled and the strapped leaf twisted. Veins and vein-lets grow beyond laminate parts of leaves giving them serrated margins. In some leaves the margins roll downward. Molybdenum deficiency leads to nitrogen deficiency. Its deficiency in poinsettias occur as yellowing of young mature leaves and may progress to scorch or burn leaf margins. This generally appears in the lower leaves, with intervein mottling and curving of the leaf. The entire edge of the leaf may dry up, the leaf is narrow and ribbon like (Whip tail).

Its deficiencies occur especially in vegetable crops when grown on very acid soils, or when it is unavailable or on soils where molybdenum is fixed by secondary soil minerals or on very well grained alkali soils. The general symptoms are: older leaves become chlorotic and margins become necrotic as nitrate accumulates. In cauliflower it causes "Whip tail" *i.e.* only mid rib develops and the lamina is absent. Its deficiencies may occur when its concentrations are less than 0.2 per cent on the dry weight basis. Both seed, soil or foliar treatment with molybdenum are useful.

The deficiency symptoms can generally be corrected by adding lime to the soil or by adding Na_2MoO_4. In leguminous plants and tomato the early symptoms of molybdenum deficiency include interveinal mottling of leaves with chlorotic areas eventually becoming puffed in appearance, marginal scorching, twisting and distortion of young leaves. The application of sodium or ammonium molybdate to the soil at 1.135kg per hectare has been shown to be satisfactory corrective (Walker,1948).

The visual deficiency symptoms of some important vegetable crops are given in Table 11.

Table 11: Deficiency Symptoms of Molybdenum in Vegetable Crops

Crop	*Symptoms*
French beans	Plants have large pale-green leaves that become mottled interveinally, and rapidly develop large irregular interveinal pale- brown scorched areas.mottled interveinally with rapidly developing brown scorched areas. Green bands remain close to mid ribs and veins even after death of affected tissue. Leaf margins wilt and leaves wither and fall. Prophylls are chlorotic, become scorched or necrotic and wither. Flowering is suppressed.
Beet root	Red veins are more conspicuous against chlorotic back ground. Leaves die in crown, which may be covered with small, deformed leaves; older leaves wilt and become necrotic. Roots show heart rot and dry rot.
Cabbage	Older leaves become mottled, bleached, scorched and coupled with irregular margins; head formation is poor. Leaves are distorted, brittle, stiff, thick mottled along margins and wilted. Leaves making up head are unattached; petioles having swellings which later become corky. In case of Savoy cabbage there is pronounced leaf cupping and membranous water-soaked margins. Veins may be purple tinted and the general colour is olive-green. Leaf angles are wide, no 'hearting', and lateral buds grow out. When secondary symptoms occur, young leaves become brown, necrotic and malformed on expansion at the margins. Rudimentary leaves turn brown and shrivel, and death of growing point follows brown exudation.
Cauliflower	"Whip tail" (leaves become strap-like with the thickening of mid rib), only mid rib develops and the lamina is absent; leaves are twisted and elongated,

Contd...

Table 11–Contd...

Crop	*Symptoms*
	Flower curds are irregular with ricy and leafy formation slightly affected plants. Leaves become strap like with the thickening of midrib. Tips develop dead patches. Cotyledons remain dark-green. Marginal limpness, wilting and scorching of leaves occur followed by leaf withering, occasional preceded by wilting of petioles. Later on there is restricted development of leaf lamina, leaving only bare midribs, producing characteristic 'Whiptail' condition. Growing point eventually die.
Cassava	Marginal scorching and rolling or curling of leaves (Chapman,1975)
Tomato	Leaves become pale with diffused marginal and interveinal yellow mottling. Margins curl upward and leaflets appear rolled. A pale brown scorching being at the tip of the apical leaflet of the oldest leaf. The distal lateral leaflets are next involved until the whole leaf is totally withered. Leaves are affected in succession and the apical leaflet scorch appears when the first or second lateral leaflets of the next older leaf are partly withered. Flowering and fruiting suppressed. Plants eventually die. The new leaves continue to become chlorotic and necrotic.
Turnip	Leaves show cupping, whiptail, chlorosis, mottling and marginal burn.
Carrot	Leaves show marginal yellowing, followed by red extending inwards. Roots have wide deep splits. Older foliage pale yellow- green. Leaf segments scorch at tips. Leaves wither and collapse.
Onion	Leaves are deep blue in colour, later the youngest leaves become conspicuously mottled, yellow and green, with distorted shrunken areas. Develop irregular, elliptic whit areas near leaf ends, which later disappear in a general break down of the affected tissues.
Pea	Leaves develop yellow or white veins, followed by small changes in interveinal areas. Growing points die. Blossoms shed. Older leaves wilt suddenly, curl up, scorch at margins and finally wither completely. Plants wither or may produce a weak basal shoot. Flowering is totally suppressed.

Contd...

Table 11–Contd...

Crop	Symptoms
Chillies	Leaf veins show decomposition and granulation, older leaves turn yellow at tips.
Radish	Terminal growing tip dies; leaves are discoloured and distorted. Roots show internal darkening. Plants pale-green with bright yellow interveinal mottling of the older leaves. Cotyledons are large and remain green for 5 to 6 weeks. Leaf margins markedly cupped. Fluid exudes on the leaf surface and water-soaked areas appear. Margins wilt, inroll and wither. Leaves wither completely except for petioles. Symptoms progress from oldest to youngest leaves until plants die. Flowers die before opening.
Roselle	Collapse of massive central portion of flower.
Cucurbits	Typical mottling and browning of leaves. Stunted plants, yellowish- green leaves; later upcurled leaf margins die off. Chlorosis and cupping in a narrow band along the margins.
Potato	Severity of symptoms dependent on tuber reserves of Mo. Symptoms usually appear in second cycle (plants raised from tubers of previous deficiency treatment). Partly or newly expanded leaves turn pale and show golden-yellow chlorosis with occasional dull white mottling. Chlorotic areas irregular and diffuse but leaf margins often dark- green. Leaf margins curl upwards, with irregular necrosis followed by general withering. Young leaflets and flowers turn brown and wither before opening, followed by the death of stems growing point. The symptoms recur in lateral branches.
Spinach	Chlorosis and severe necrosis of successive leaves followed by death of plants. Mulder (1954) reported pale-yellow leaves, marginal cupping and necrosis.
Broad bean	Lower leaves pale, margins and interveinal areas become grey and wilted or collapsed. Dark-brown or black scorching rapidly follows, while blackening appears in younger leaves. Stems become almost totally defoliated. Flowering and seed production are almost completely suppressed.

Contd...

Table 11–Contd...

Crop	*Symptoms*
Runner beans	Leaves tend to be small, uniformly pale yellow-green or yellow and finally bleached with scorched up cupped margins. Older leaves are affected first and wither before falling. Stem elongation and flower and seed production suppressed. Plants wither.

Copper

About 70 per cent of copper (Cu) in plant tissue is found in organelles of chloroplasts (Mengel and Kirkby,1978) and functions as a component of phenolases, lactase, ascorbic acid oxidase, uricase, diamine oxidase, cytochrome oxidase and plastocyanin (Evans and Sorger,1966). It participates in protein and carbohydrate metabolism and nitrogen fixation. It is a constituent of chloroplast protein, plastocyanin and part of many enzymes such as cytochrome oxidase, ascorbic acid oxidase and polyphenol oxidase. It is involved in the de-saturation and hydroxylation of fatty acids. Its deficiency causes the young leaves to curl, and the entire leaf becomes cupped and leaflet margins turn up-wards. The plants become stunted, rosette, and there is interveinal crinkling and marginal wilting which occurs due to weakness of cell wall, but not due to water stress. Irregular leaf lets with marginal necrosis, mild chlorosis and small yellow- white spots develop on the foliage. Bronzing and necrosis of the outer edges of the leaflets occur if deficiency is prolonged. Copper deficiency also reduces root growth more then shoot growth, creating an unfavourable shoot: root ratio. There is decrease in flower pigmentation. It occurs due to low natural copper soil content or low availability. Normal plants should contain 6ppm copper on dry weight basis (Maynard, 1979).

Role in Plant

- ☆ It participates in protein and carbohydrate metabolism and nitrogen fixation
- ☆ Copper is an essential co-enzyme for oxidase enzymes catalyzing the oxidation of phenols and ascorbic acid.
- ☆ In photosynthesis, it is a part of the chloroplast enzyme plastocyanin in the elcctron transport-system, between photosystems 1 & 11.
- ☆ Most of copper in plants is found in the organelles of chloroplast.
- ☆ It is part of several oxidases, such as ascorbic acid oxidase and polyphenol oxidase.
- ☆ Copper is involved in de-saturation and hydroxylation of fatty acids.

Deficiency

Copper is immobile so deficiency is first visible in new leaves and shoots. They look bleached (apical bleaching), turning and looking dry and soft. The leaves located immediately below the tip are often unable to hold themselves up right. Deficiency causes severe stunting of plants, with terminal leaves becoming small (often only one-fifth or less of normal size). Terminals die as the stunting advances, and multiple budding occurs immediately below dead terminals. These buds will also die and often multiply, buds will develop on each break, giving the plant a "witches broom" appearance (Dickey,1977). Growth of tomato plants having deficiency of copper remains stunted, foliage turns dark bluish green, leaves curl, and there is chlorosis and absence of flower formation. Root development is poor.

In case of onions a thin and yellow scales are produced, which lack solidity and firmness. Oats are prone to copper deficiency, while soybean is highly tolerant to its deficiency.

Its deficiency causes chlorosis, necrosis and wilting of young leaves and the number of irregularities caused due to its deficiencies are described in Table 12.

Table 12: Deficiency Symptoms of Copper in Vegetable Crops

Crop	*Symptoms*
Tomato	Growth of shoots is stunted and root development is exceedingly poor. Other symptoms seen are dark bluish-green colour of the foliage, leaves curl upward and inward and reduction in size of leaves and a lack of firmness in leaves and stems. There is wilting of the entire top of the plant. At later stages necrotic areas may appear in the leaf tissue. There is absence of flower formation.
Onion	Scales are abnormally thin and are of pale-yellow colour. Cu-deficient onions lack solidity and firmness. The yellow scale colour intensify & scale thickness increases by adding 250 – 500kg per hectare of powdered cooper sulphate to the soil and dishing it before planting.
Brinjal	Plants are severely stunted and chlorotic with green spots on the leaves and marked yellowing and browning at the leaf tips.
Peas	The noticeable symptom is the limpness of the top few inches of the main stem; considerable drooping occurs, and it is unable to support itself in an erect position. Apart from this, vegetative growth appears normal and no characteristic leaf symptoms are observed. Only a few flowers are formed, no seed is formed and the plants die earlier than those receiving copper. Under field conditions the decreased seed production is the most important feature of the disease in peas.
Pepper	Plants are severely stunted and become chlorotic with leaves cupped, crinkled and twisted.

Contd...

Table 12–Contd...

Crop	Symptoms
Carrots	Growth is severely checked. Plants are highly stunted with very small roots.
Cabbage	Plants are chlorotic stunted, and do not attain enough growth to produce heads.

Its deficiency causes the young leaves to curl, and the entire leaf becomes cupped and leaflet margins turn upwards. The plants become stunted, rosette, inter-veinal crinkling and marginal wilting occurs due to weakness of cell wall, but not due to water stress. Irregular leaflets with marginal necrosis, mild chlorosis and small yellow- white spots on the foliage. Bronzing and necrosis of the outer edges of the leaflet occurs if deficiency is prolonged. Copper deficiency also reduced root growth more than shoot growth creating an unfavourable shoot: root ratio. The pigment in flower decreases. Its deficiency first causes severe stunting of plants with terminal leaves becoming small often (only one fifth or less of normal size). Terminals die as the stunting advances and multiple budding occurs immediately below dead terminals. These buds will also die and often multiple buds will develop on each break, giving the plant a witches broom appearance (Dickey, 1977). Its deficiency can be corrected by incorporation of copper sulphate in the fertilizer or spraying with a copper fungicide.

Manganese

The primary role of manganese (Mn) in plants is that of an enzyme activator for oxidation- reduction reactions. It activates such enzymes as catalases, dehydrogenases, decarboxylases, kinases, oxylases, phosphatases, peroxidases and other enzymes activated nonspecifically by divalent cations (Evans and Sorger,1966). These enzymes control such metabolic processes as synthesis of fatty acids and nucleotides, activate the breakdown of carbohydrates in respiration and phos-phorylation reactions, function in nitrate reduction and vitamin synthesis and control and are associated with the oxygen- evolving system of photo-synthesis by effecting photo-oxidation of water (Gerretson,1950). Besides being activator of several enzymes, especially those involved in fatty acid and nucleotide synthesis is also essential in respiration and photosynthesis. It also activates indoleacetic acid (IAA) oxidase, which results in less IAA concentration in tissue.

Manganese deficiency is reported to occur in soils having pH above 6.8, because of the insoluble manganic due to oxidation of soluble manganous. It imparts oxidation reduction process, photosynthesis, oxygen evolution, activates IAA enzyme complex phosphokinase and phosphotransferases. Several enzymes activated by Mn and Mg can also function, but the photosynthesis and regulation of IAA are the highly Mn specific activity. With Manganese, plants accumulates H_2O_2 causing cell damage. The calcareous soils are Mn deficient owing to immobilization as insoluble MnO_2 at high pH.

Role in Plant

- It is essential for the synthesis of chlorophyll, as intense chlorosis results under its severe deficiency.
- It also activates certain enzymes especially those involved in fatty acid and nucleotide synthesis.
- It is essential in respiration and photosynthesis.
- It also activates indoleacetic acid (IAA) oxidase, which results in less IAA concentration in tissue.

- ☆ Acts as a bridge for ATP & enzyme complex phosphokinase and phosphotransferases
- ☆ Functions in nitrate reduction & vitamin synthesis and control

Deficiency

It is not mobile in plants, therefore deficiency symptoms appear on young tissue first and are similar to Fe deficiency except that the persistent bands of green along veins and vein lets are broader and include tissue immediately adjacent to veins and vein-lets. Interveinal chlorosis is not as severe as with Fe deficiency and seldom progresses past the green- yellow state. There is little to no reduction in leaf size (Dickey,1977). Chlorotic patches appear allover the leaf and develop into intervein necrosis. Soybean, pea and sugar beet are susceptible to manganese deficiency. In onion there is curling of leaves, slow growth, light colour of the foliage, delayed bulbing and thick necks.

Its deficiency causes chlorosis followed by necrosis. Deficiency symptoms first appear on the young leaves of interveinal areas and they develop when the leaf manganese concentration is less than 50ppm on dry matter basis.

In beans, manganese deficiency occurs in soils which are nearly neutral or alkaline, as it becomes unavailable. The trifoliate leaves show progressive loss of green colour, begin with a faint mottle, followed by appearance of small necrotic brown spots and finally of a fairly uniform golden-yellow colour. Growth is retarded and each leaf is smaller and more chloratic than the previous one. Finally the buds die and leaves become brown and withered. The deficiency can be prevented by applying 62.5–125 kg of manganese sulphate/ha with fertilizers. It may be controlled on the growing crop if two or three applications of the same material are applied at the rate of 4.54 kg/568 lit/ha. Spraying of manganese sulphate with 0.1 to 0.2 per cent may produce beneficial results.

Like Fe, Mn is relatively immobile and preferentially translocated to young or meristematic tissues. These parts can not depend on transfer from older leaves and hence are the first to show Mn deficiency as lesions on younger leaves. When soil is known to be deficient in manganese, the sulphate should be added with the fertilizer at the rate of about 113.5 kg per hectare. Spray with a solution of manganese sulphate can be used.

The visual deficiency symptoms of some important vegetable crops are given in Table 13.

Table 13: Deficiency Symptoms of Manganese in Vegetable Crops

Crop	*Symptoms*
Beans	Interveinal chlorosis. In severe cases the entire leaf turns yellow to yellowish- green, and the veins remain green and stand out in marked contrast. Small brown spots appear on each side of the midrib & veins till the leaves become completely yellow. The seeds may have 'marsh spot'.
Beet root	Leaves are chlorotic between veins, with erect growth, margins are curled towards upper surfaces; red and purple tinting appears.
Cabbage	Leaves are smaller and yellower than normal, and are marked by yellow mottling between veins.
Cassava	Produces interveinal chlorosis of young recently expanded leaves.
Potato	Reduction in inter-node length and leaf chlorosis. Later numerous small brown patches develop along the veins and these areas increase both in size & number. Interveinal light green chlorotic areas on upper leaves.Top parts of the plant chlorotic, leaves turning pale, losing their luster, being smaller and curled up;

Contd...

Table 13–Contd...

Crop	Symptoms
	numerous necrotic spots appearing along the veins; in acute deficiency much browning and yellowing of plants, especially young leaves.
Pea	The plants may appear quite healthy or in severe deficiency, may have slightly chlorotic foliage. Young tendrils and internodes at the top show brownish discolouration. The youngest leaves may fail to expand. Slightly older leaves acquire characteristic mottling due to green veins and yellow interveinal areas. The lower leaves retain green colour. Later, growing tips and younger leaves may die. The seeds show a very characteristic condition known as 'marsh spot disease.' Small brown speaks or larger circular brown areas are seen on the inner flat surfaces of the cotyledons. The areas may become hollowed out in very severe deficiency.
Tomato	Shoot growth varies from normal to severe, stunting; leaves near shoot tips are small, rolled forward, and some what chlorotic; most varieties show small dark brown spots along veins or distributed sporadically on younger leaves. In areas farthest from the major veins the leaf colour turning light green and finally yellow; yellowing gradually extending into larger areas; the veins remaining green and a mottled appearance develops. Necrosis sets in as small pin-points at the center of the yellow areas; little or no blossoming; no fruit formation.
Onion	Curling of leaves & chlorotic streaking of leaves. Growth is slow. Light colour of foliage; delayed bulbing and thick necks. Apply manganese sulphate or elemental sulphur to the soil.
Cucumber	Webs of leaves change from green to yellowish white, while regions along the veins & midribs retain green colour. Stems & leaves remain small in size & slender. Frequently blossom buds turn yellow.
Pepper	Plant growth is stunted and yield is reduced.
Okra	Reduction in inter node length, leaves show chlorosis

Contd...

Table 13–Contd...

Crop	Symptoms
Spinach	Cholorosis starts in the growing tips and gradually extends throughout the plant. The web of the leaf turns pale-green to golden yellow. The green colour persists longest along the main veins. Later on white dead areas may appear between the leaf veins.
Sweet potato	The leaves develop slight cholorosis with the vein network remaining green. In case of a severe deficiency the back ground lightens to a pale yellowish-green. Younger leaves are affected first. Vine growth becomes some what limited. Sweet potato production is reduced but the size is not much influenced.
Brassica crops	The symptoms first appear as an interveinal chlorotic marbling. Under severe deficiency the entire leaves may be practically bleached, only the veins remaining green (kale) or some necrotic tissue may develop in the mottled tissue, when it takes on a dull brownish appearance (Savoy cabbage). Plant growth is stunted and yield is reduced.

Zinc

In alkaline soils zinc sometimes becomes unavailable. The leaves become distorted, thick and tough and growth of the plant becomes retarded in mid season. Yellow brown or whitish chlorotic spots develop between the larger veins and the leaves fall pre-maturely. Deficiency symptoms in vegetable crops are mostly species related. In general, its deficiency results in shortened internodes and chlorotic areas on older leaves or may appear in younger plants also. Shortened internodes may be due to non-availability of indoleacetic acid, as zinc is essential for the synthesis of tryptophane, a precursor of IAA. Normal healthy leaves contain 20ppm zinc. Zinc sulphate added to the fertilizer at 12.5 to 25kg/ha or a spraying to the growing crop of beans with 2.5kg in 568 lit of water per hectare are usual remedy. Zinc deficiency is also aggravated by phosphorus application. Zinc plays a role in the carbonic anhydrase chloroplast enzyme system, which control CO_2 fixation in photosynthesis. In tomato zinc is required for the formation of ascorbic acid in fruits, but its higher levels may cause a reduction in carbohydrates.

In onion there is corkscrewing of the leaves or in mild cases a general outward bending of the leaves which give them a flattened appearance. There is irregular orange splotching of the older leaves and a very faint interveinal chlorosis of the younger leaves.

Role in Plant

- It helps in carbohydrate metabolism and probably in the turnover of organic phosphorus compounds.
- The production of auxins is also retarded due to its deficiency.
- It is also a constituent of carbonic anhydrase, which catalyzes the reaction $H_2CO_3 = H_2O + CO_2$
- It plays a primary role in plant metabolism in the synthesis of tryptophan, a precursor of indoleacetic acid.
- Zinc plays a role in the carbonic anhydrase chloroplast enzyme system, which controls CO_2 fixation in photosynthesis.
- Aids in maintaining the structure of polysomes.

Deficiency

- Plants deficient in zinc are low in tryptophan and IAA and exhibit small leaves and early abscission.

- ☆ Causes reduction of RNA synthesis and ribosome stability.
- ☆ Zinc deficient plants have decreased water content and increased osmotic density.

It is non mobile in plant tissue, therefore deficiency systems appear in young tissue first. Symptoms appear as small leaves that decrease faster in width than in length and appear elongate. There is a rapid reduction in internodal length, so that terminal leaves "rosette." Veins and vein lets are reduced more than the laminate portion of leaves to produce a cupping and twisting of affected leaves. Often one side of a leaf will be much more reduced than the other, bending the leaf into the shape of a "C." Interveinal chlorosis occur and rapidly growing patches occupy the spaces between the veins, some times invading the nerves. Because auxin synthesis declines, the internodes are shorter and the leaves may become small and thickened. In the final stages, the leaves may necrose on the edges and at the tip.

When it is deficient, young leaves turn yellow, then white and brown. In tomato leaves become smaller, chlorotic, and the whole plant is dwarfed with small leaves curling inward, where as in onion leaves turn yellow. In case of Bakers garlic whole plant turns yellow and develops a rosette form with poor root growth. Beans and lima beans are quite susceptible to zinc deficiency; potato, tomatoes and onions are some what sensitive to zinc deficiency whereas, peas, asparagus and carrot are insensitive to zinc deficiency. In lettuce zinc deficiency, there is stunted growth, marginal purpling, rupture of margin of mature leaves and overall chlorosis to interveinal chlorosis. Sarin and Saxena (1965) reported that okra plants when subjected to Zn deficiency resulted in a depression in leaf production, the leaf size appeared within three weeks and a leaf molting developed about two weeks later. The stem diameter was reduced, and the rate of increase in dry weight and the rate of P uptake were also depressed within three weeks, whereas ribonuclease activity increased. Addition of organic matter to soil or growing green manuring crops frequently improves crops subject to zinc deficiency. In soybean small leaf size is the first visible symptom, followed by chlorosis in the youngest leaves.

Common symptoms of zinc deficiency are the interveinal mottling, light colour of the chloroic areas, necrosis or death of the affected areas on the leaves.

The visual deficiency symptoms of some important vegetable crops are given in Table 14.

Table 14: Deficiency Symptoms of Zinc in Vegetable Crops

Crop	*Symptoms*
Broad bean	Leaves and flower buds shed; seed pods fail to develop.
Bean	Leaves and flower buds shed.
Pea	Lower leaves necrotic at margins and tips; stems stiff and erect; flowers none
Potato	Plants stunted. First symptoms appear on middle leaves in the form of greyish brown to bronze coloured irregular spots, which finally develop on all the leaves. The affected tissue becomes sunken and eventually dies. Severe cases brownish spots develop on leaf petioles and stems. The young leaves roll upward in such a way that the apical growth resembles that of fern and hence the deficiency symptoms is also known as' fern leaf.'

Contd...

Table 14–Contd...

Crop	*Symptoms*
Squash	Leaves mottled with necrotic areas.
Sugar- beet	Leaves necrotic with brown to greyish spots, tips withering, only petioles green
Tomato	Leaves mottled, curling, distortion and necrotic spotting of leaflets. leaflets small, midrib shortened.
Onion	Corkscrewing of the leaves or in mild cases a general outward bending of the leaves which gives them a flattened appearance. Irregular orange splotching of the older leaves and a very faint interveinal chlorosis of the younger leaves.
Soybean	Small leaf size initially followed by chlorolysis in the youngest leaves. Acute deficiency showed initial chlorosis in the interveinous parenchyma, then reduced leaf growth & finally dieback.

Iron

It is the 4th most abundant element in the earth and soil, still its deficiency is most widespread in the world mainly due to its availability in root zone rather abundance. Iron is a component of cytochrome oxidase, ferredoxin protein and several enzyme systems, precursor of chlorophyll and in mitochondrial respiration. It is involved in nitrate and sulphate reductase, nitrogen assimilation, energy (NADP) production and of the N_2 fixation enzyme nitrogenase. It is involved in activation of several enzyme systems, including fumeric hydrogenase, catalase, oxidase and the cytochromes (Evans,1959), and is a component of ferredoxins, which are nonheme, Fe-S cluster proteins involved in N fixation, photosynthesis, and electron transfers.

Role in Plants

- It is a constituent of electron transport enzymes e.g; the cytochromes & ferredoxin,which are active in photosynthesis and in mitochondrial respiration.
- It is a constituent of the enzymes catalase and peroxidase, which catalyze the breakdown of H_2O_2 into H_2O & O_2 preventing H_2O_2 toxicity.
- Although iron is not a part of the chlorophyll molecule, but affects chlorophyll levels because it must be present for chloroplast ultra–structure formation.
- Chlorophyll synthesis although it is not a part of its moecule.
- Enzyme activation (acid phosphates, aldolase etc).
- It is incorporated in several important oxidation enzymes which are involved in the process of respiration

Deficiency

Typical Fe deficiency begins as a mild interveinal chlorosis with veins and vein lets remaining green. The interveinal areas progressively turn from green- yellow to yellow- green to yellow; and finally to a cream colour. By the time the deficiency is so severe that leaves have become cream coloured, veins and vein lets have lost their persistent green colour and the entire leaf is affected. These leaves are small and frequently abscise (Joiner and Waters,1970). There is reduced growth and interveinal yellowing in spinach and potatoes (Tokutaro Inden, 1975). In brinjal, yellow spots are usually found to occur on the fifth or sixth leaf. Iron functions in plants in many ways, but a lack of iron in growing medium most often evidenced by a yellowing of the leaves. Commonly referred to as iron chlorosis. Iron chlorosis involves a reduction in the chlorophyll content of leaves. The resultant decrease in photosynthesis directly affect plant growth and development and reduces the productivity for economic uses by human being. Its deficiency may restrict nitrogen fixation by limiting the host plant growth in general and active functioning of rhizobium in particular.

The symptomology of iron deficiency (Table 15) is usually manifested as an interveinal chlorosis of young leaves, while the veins remain green-hence name iron deficiency chlorosis. The expression of the symptoms in young leaves is due to the inability to redistribute iron within the plant. The first symptom of its deficiency occurs in younger leaves as yellow mottling of interveinal areas. The affected tissues becomes lighter in colour and entire leaf becomes pale yellow to almost white with little necrosis. In potato, slight chlorosis in the young leaves occur due to iron deficiency which regularly spreads over the blades. The points and margins of leaflets retain the green colour for a long time. As both the green and yellow pigments are affected, the discoloured tissue becomes pale yellow and in extreme cases almost pure white. The iron deficiency can be controlled by spraying plants with week solution of ferrous sulphate (0.25 per cent). Iron deficiency lead to interveinal chlorosis of young leaves followed by yellowing and pre-mature shedding. The malady may be corrected by watering with a weak solution of iron sulphate.

Deficiency Symptoms

- ✰ Reduces the number and size of chloroplasts.

- ☆ Grana and lamella of the chloroplasts are reduced in Fe deficient maize.
- ☆ Chlorosis
- ☆ Leaves small & frequently abscise.
- ☆ Yellowing of leaves
- ☆ Loss of leaves

The general symptoms of iron deficiency occur in younger leaves as yellow mottling of the interveinal areas. The affected tissues become lighter in colour and entire leaf becomes pale to almost white with little necrosis.

The visual deficiency symptoms of some important vegetable crops are given in Table 15.

Table 15: Deficiency Symptoms of Iron in Vegetable Crops

Crop	*Symptoms*
Tomato	Leaves near tips of plants show chlorotic mottling, the chlorosis being most pronounced near the mid rib and towards the bases of the leaflets. Little necrosis or drying of leaf tissues may occur. While there may be chlorolysis with little or no necrosis.
Potato	Plants grown in water culture show slight chlorosis of young leaves rather regularly spread over the leaf-blade. The points and margins of leaflets keep their green colour longest. Discoloured tissues become clear pale- yellow and in extreme cases almost pure white. Chlorotic tissue is curved in an upward direction. Leaves developed before appearance of symptoms remain normal green. These symptoms have not been reported so far from fields, where the plants are quite resistant.
Brassica crops	Chlorotic 'marbling' of leaves and occasionally leaves are completely bleached

Chlorine

The micronutrient most recently confirmed as essential for growth of higher plants is chlorine (Hajrasuliha,1980), although there are still many speculations in the literature relative to its exact physiological roles. Chlorine (Cl) increases the effectiveness of cylic and non-cyclic photophosphorylation and activates electron transfer for liberation of oxygen (Arnon,1959).Chlorine is involved in stomatal function. It is not a constituent of any known plant metabolism but was found essential for the evolution of oxygen in photosynthesis.

Chlorine along with Mn, is required for oxygen evolution in photosynthesis. It is involved in osmosis (movement of water or solutes in cells), ionic balance necessary to take up mineral elements. Chlorine ion is ubiquitous in nature and highly soluble and plant extract it from the soil more rapidly than many other ions, and accumulates excess of chlorine than their requirement. In plant chlorine is a major counter ion maintaining electrical neutrality across membrane and osmotically active solutes in the vacuole. It is also required for cell division in both leaves and shoots.

Role in Plant

- ☆ Nitrogen metabolism
- ☆ Cell division
- ☆ Photosynthesis
- ☆ Chlorine increases the effectiveness of cyclic & non-cyclic-Photophosphorylation.
- ☆ It activates electron transfer for liberation of oxygen.

☆ It is involved in stomatal functions

Deficiencies

Plants deficient in chlorine exhibits reduced growth, the symptoms first appear on young actively growing leaves as wilting of the leaf tips and normal chlorosis, stubby roots yellowing and bronzing. Most soils contain sufficient chlorine and receive enough of it in fertilizer especially potassium thus, the deficiency is rarely observed (Table 16).

Deficiency symptoms appear first as wilted leaves that subsequently become chlorotic and bronze coloured. The contribution of chlorine to reduction of lodging has been a frequent question, since observation of reduced lodging were made when fertilizer KCl was used.

Table 16: Deficiency Symptoms of Chlorine in Vegetable Crops

Crop	*Symptoms*
Potato	Slight chlorosis in the young leaves, which regularly spreads over the blades. Points & margins of leaflets retain the green colour for a long time. Since both the green & yellow pigments are affected, the discoloured tissue becomes pale yellow & in extreme cases almost pure white.
Tomato	Leaves near tips of plant show chlorotic mottling, the chlorosis being most pronounced near the mid rib & towards the bases of the leaflets. Little necrosis or drying of leaf tissue may occur.
Brassica crops	Chlorotic 'marbling' of leaves, occasionally leaves are completely bleached.

☆ Reduced shoot and root growth

☆ Leaves wrinkled

☆ Chlorosis of younger leaves

☆ Wilting of the plant

Excess

☆ Premature yellowing of leaves

☆ Burning of leaf tips and margins

☆ Bronzing and abscission of leaves

Correction of chlorine deficiency is seldom, if ever, needed, since adequate amounts can come from the air, rains and from animal urine ad perspiration.

Table 17: Mineral Related Disorders

Mineral	*Crop*	*Disorder*
Calcium	French Bean	Hypocotyl necrosis
	Brussels sprout	Internal browning
	Cabbage	Internal tip burn
	Chinese cabbage	Internal tip burn
	Carrot	Cavity spot, cracking
	Celery	Black heart
	Chicory	Black heart, tip burn
	Escarole	Brown heart, tip burn
	Lettuce	Tip burn
	Taro	Metsubre
	Parsnip	Cavity spot
	Peppers	Blossom-end rot
	Potato	Sprout failure, tip burn, black necrotic spots, ageing of tubers
	Stranben	Leaf tip burn
Strawberry	Leaf tip burn	
	Tomato	Blossom-end rot, black seed, cracking, ripening disorders
	Watermelon	Blossom-end rot

Contd...

Table 17–Contd...

Mineral	Crop	Disorder
Boron	Celery	Cracked stem/brown checking
	Brinjal	Internal chlorosis
	Turnip	Brown heart
	Sugar Beet	Heart rot
	Potato	Leaf roll
	Celery	Cracked stem
	Cauliflower	Brown curd
	Root crops	Brown heart
Magnesium	Celery	Chlorosis
	Tomato	Vascular browning
Potassium	Potato	Internal bruising, Hollow heart
Boron	Beet root	Brown heart. Black spot
	Carrot	Splitting of carrots
	Cauliflower	Browning
Potassium	Tomato	Blotchy ripening
Phosphorus (Excess)	Celery	Pencil strip
Molybdenum	Cauliflower, broccoli	Whip tail, die back & internal chlorosis
Magnesium	Cauliflower	Chlorosis
Nitrogen (low)	Cauliflower	Buttoning
(Excess)	Carrot	Splitting
	Cauliflower	Ricyness & Hollow stem (due to excess N)
	Garlic	Bulb Sprouting
Potassium	Potato	Hollow heart
Potassium	Sweet potato	Cracks (checks cracking)
Ethylene	Sweet potato	Russet spotting
Nitrogen	Cauliflower	Hollow stem
Ethylene	Carrot	Bitterness

Contd...

Table 17–Contd...

Mineral	*Crop*	*Disorder*
Boron	Cauliflower	Browning of heart
	Tomato	Cracking, pitted and cracky areas, uneven fruits
Boron	Beet	Internal black spot
Zinc	Tomato (helps)	Formation of ascorbic acid
	Soybean	Small leaf size.
	Onion	Corkscrewing of leaves & Splotching of older leaves
	Potato	Fern leaf
Boron	Broccoli, cauliflower	Hollow stem
Ethylene	Lettuce	Russet spotting (brown spots)
Manganese	Beet root	Yellow blotches between leaf veins and leaves
	Onion	Curling of leaves
Nitrogen	Potato	Hollow heart
Calcium	Potato	Internal rust spot
Magnesium	Cauliflower	Cholorosis (in acid soils)
CO_2 (excess)	Lettuce	Brown stain

Toxicity of Minerals

The range of difference between deficiency and toxicity of essential nutrients is often small and depends on a variety of factors in the environment that affect their availability to plants. Toxic levels of zinc, copper and nickel occur frequently in soil. Toxic concentrations of lead, cobalt, beryllium and cadmium can occur but only under unusual conditions (Foy *et al.,* 1978). Aluminum toxicity symptoms occur as stunting, small dark green leaves and late maturity with a purpling of the leaves and stems, yellowing and death of the leaf tips. There is induced calcium deficiency that results in rotting of young leaves and collapse of growing points or petioles. Roots become brittle and stubby with a spatulate shape and root tips with a coralloid appearance. The toxic effects of aluminum are counteracted by boron (Belvins and Lukaszewski, 1998). Fluoride injury is often characterized by tip burn of leaves. Ammonium toxicity shows stunting and chlorosis in most of the plants (Gerendas, *et al.*, 1997). Carbon dioxide toxicity has also been reported in some cultivars of tomato, radish and beans. There is delayed development, leaf yellowing and crumbling to high levels of carbon dioxide.

Air Pollution Damage

Vegetables are injured by exposure to high concentration or a prolonged exposure to a lower concentration of an air pollutants (Table 18). Damage caused by the air pollutants is frequently confused with nutrient deficiency symptoms or pest damage. When plants are injured by air pollutants, symptoms characteristic of that specific pollutant usually develop. Pollutants on contact with plant tissue, changes itself and the damage symptoms are frequently the only remaining evidence of air pollution (Table 18).

Ozone and PAN (Peroxyacetyl Nitrate)

Ozone is extremely photo-toxic. It is presently considered among the major air pollutants not only in urban areas, but also in rural and remote sites. Realistic fumigation studies have shown a range of metabolic responses, including visible injury, reduction in biomass, alteration of assimilation rate, acceleration of the senescence process, variation in primary and secondary metabolism (Manning *et al.*, 2002). Vegetables which are cultivated in open fields, where they may be exposed to pollutants, such as O_3, that could induce modification in metabolism, with possible adverse effect on their nutritional and pharmaceutical properties. O_3 is a strong oxidant that can interact with constituents of the apoplast and generate reactive oxygen species (ROS), interfering with essential oil and volatile emitted compounds. Ozone oxidizes plant surfaces and tissue and affects many important physiological processes. It is particularly injurious to membranes and notably inhibits photosynthesis and biomass accumulation, suppresses phloem loading, reduces carbon allocation to the roots. These actions result in ozone being rated as the pollutant of major concern in crop production. Ozone depletion has implication for climate change; it contribute to global warming and intercepts ultraviolet-B radiation.

Ozone damage appears as very small, irregular shaped spots on the upper leaf surface. These spots are dark brown to black and after 24 hours become light tan and then white after some days (Hindawi,1970). Recently matured leaves are most susceptible, where as very young and old leaves are most resistant. It is the leaf tip and the margins which are damaged the most. Onion and melons among all vegetables are affected the most.

While as peroxyacetyl nitrate affects under side of newly matured leaves and the effected area becomes bronzed, glazed or silvery in appearance. PAN injury develops on only three or four rapidly expanding leaves of sensitive plants. Very young and mature leaves are highly resistant and pale green to white areas may appear on the leaf surface.

Nitrogen Dioxide

It is produced by the combustion of coal, fuel, natural gas and gasoline used in power- generation operations. Vegetable cultivars vary in their susceptibility to nitrogen dioxide and the injury symptoms appear as irregular white or brown collapsed lesions on leaf tissue between the veins and near the leaf margins. The concentration of nitrogen dioxide influences plant injury more than the duration of exposure. Damage can be easily noticed in vegetable areas.

Table 18: Relative Susceptibility of Selected Vegetables to Damage by Some Air Pollutants

Air Pollutant	*Most Susceptible*	*Intermediate*	*Least Susceptible*
Ozone	Bean	Carrot	Beet
	Broccoli	Endive	Cucumber
	Muskmelon	Parsnip	Lettuce
	Onion	Turnip	–
	Potato	–	–
	Radish	–	–
	Spinach	–	–
	Sweet corn	–	–
	Tomato	–	–
Peroxyacetyl	Bean	Beat	Broccoli
Nitrate (PAN)	Celery	Carrot	Cabbage
	Chard	Sweet corn	Cauliflower
	Endive	–	Cucumber
	Lettuce	–	Onion
	Pepper	–	Radish
	Spinach	–	Squash
	Tomato	–	–

Contd...

Table 18–Contd...

Air Pollutant	*Most Susceptible*	*Intermediate*	*Least Susceptible*
Nitrogen dioxide	Lettuce	Dandelion	Asparagus
	Mustard	–	Bean
Fluoride	Sweet corn	–	Asparagus
	–	–	Squash
			Tomato
Chlorine	Onion	Bean	Brinjal
	Radish	Cucumber	Pepper
	Sweet corn	Pea	--
	–	Dandelion	–
	–	Squash	–
	–	Tomato	–
Sulphur dioxide	Beet	Cabbage	Cucumber
	Broccoli	Pea	Onion
	Bean	Tomato	Sweet corn
	Brussels sprouts	–	–
	Carrot	–	–
	Chard	–	–
	Endive	–	–
	Lettuce	–	–
	Okra	–	–
	Pepper	–	–
	Pumpkin	–	–
	Radish	–	–
	Rhubarb	–	–
	Spinach	–	–
	Squash	–	–
	Sweet potato	–	–
	Turnip	–	–

Source: Jacobson and Hill (1970) and Hindawi (1970).

Sulphur Dioxide

Acute injury is characterized by dead tissue between the veins or on the margins of leaves. The dead tissue may be white, silver, tan brown, orange, or red in colour, depending upon the plant species and time of the year exposed. While chronic injury is marked by brownish-red, turgid, or bleached white areas on the blade of the leaf. Fully expanded leaves are most sensitive, while young leaves seldom show damage (Hindawi, 1970).

Fertilizer Application Techniques

Nitrogen fertilizer applied in large amounts prior to planting or early during crop growth is exposed to the risk of leaching or di-nitrification. This risk can be reduced to an great extent by splitting up nitrogen fertilization in to small portions.

The use of granulated nitrogen fertilizers for top dressing may be restricted to early growth stages, depending upon the crop. The injection of nitrogen fertilizer in the irrigation water, however, allow top dressing the entire growth cycle. This technique which is known as fertigation, allows precise timing of nitrogen in very short time intervals (weeks, days or even shorter) according to the demand of the crop.

Benefits from splitting nitrogen application only occur if losses are prevented and/or net mineralization deviates from the average. Potentially, beneficial effects can be expected (1) on soil having low water capacity (2) with shallower root crops, (3) in regions of heavy- rainfall with risk of nitrate leaching.

For P that is relatively immobile in the soil, it is well established that banding of P fertilizers can improve P efficiency. Placement of a nitrogen fertilizer doses on the plant could improve spatial availability for crops at large planting distances during early growth stages when plant roots have not yet penetrated the whole potentially rootable soil volume.

Crop Species Sensitive to Either Deficient or Excessive Levels of the Micronutrients

Micronutrient	*Sensitive Crop*	*Sensitive to Excess*
Boron	Legumes, Brassica (Cabbage & relatives) beet, Sugar beet	Potato, tomato, cucumber
Chlorine	Potato, sugar beet, lettuce, carrot, cabbage	Navy bean, pea, onion
Copper	Spinach, onion, watermelon	Legume, spinach
Iron	Tomato, spinach, brassica	–
Manganese	Legumes, sugar beet, potato	Legumes, potato, cabbage
Molybdenum	Brassica (Cabbage & relatives), legumes	Pea, green bean
Zinc	Legumes	Spinach

Source: Benton, & Jones 1998.

Time and Method of Fertilizer Application

To obtain the maximum benefit from fertilizers, it is most important that fertilizers are applied at the proper time and at proper place. The fertilizers to be applied possess qualities with regard to solubility in water and movement into the soil solution. Similarly, soils are of different nature–sandy- clayey. The nature of the soil governs the movement of applied fertilizers. The requirement of plants for different plant nutrients varies in relation to their stage of growth. For example nitrogen is absorbed by the plants throughout the growth period, while phosphorus is absorbed at a faster rate during the early growth period. Therefore, time and method of fertilizer application varies in relation to nature of fertilizer, soil type, nutrient requirement and the nature of the vegetable crop.

Nitrogen is required throughout the crop growth and is absorbed by the plants at the same rate as that of its growth. Accordingly, nitrogen is taken up by the plant slowly in the beginning, rapidly during the grand growth period and again slowly as it nears maturity. Therefore, nature of the crop *i.e.* duration of the crop should also be kept in view while applying the fertilizers. Short duration crop like beans etc. will require only basal dose of nitrogenous fertilizers.

Nitrogenous fertilizers are soluble in water, move rapidly in all directions within the soil, but are easily lost through leaching. Therefore, it is better not to apply too much nitrogenous fertilizers at one time but apply in split doses throughout the growth period. As it will supply nitrogen to the growing plants during the entire growth period and the plants will not suffer from nitrogen deficiency.

While as phosphorus is required during the early root development and early plant growth. As such, crop plants utilize 2/3 of the total requirement of phosphorus when the plants accumulate 1/3 of their dry weight. All phosphatic fertilizers release phosphorus for plant growth slowly. As on application of superphosphate to the soil, water soluble P_2O_5 becomes immediately slightly insoluble or is converted into di-calcium phosphate or citrate soluble P_2O_5. In this form, phosphorus becomes available to plants slowly.

On the other hand potash behaves partly like nitrogen and partly like phosphorus. But potassic fertilizer, like phosphateic fertilizers, become available slowly. Therefore, it is always advisable to apply the entire quantity of the potash at sowing time.

Methods of Applying Fertilizers

Broad Casting

In broadcasting, the fertilizer is spread over the entire soil area to be treated, with the main objective of distributing the whole quantity of fertilizer evenly and uniformly. This may be done immediately before planting, or when the crop is standing.

Topdressing

Nitrogenous fertilizers containing nitrate nitrogen, like sodium nitrate, calcium ammonium nitrate, ammonium sulphate-nitrate, ammonium nitrate and di-ammonium phosphate (DAP) are top dressed mostly to closely spaced vegetable crops especially when the objective of supplying nitrogen in readily available form to growing plants. One to several top dressings of nitrogenous fertilizers may be made to provide nitrogen at the time of greatest need of the crops. But utmost care need to be taken that leaves are not wet at the time of top dressing, for it may burn or scorch the leaves. This injury is greater with nitrogenous and potassic fertilizers than with phosphatic fertilizers.

Band Placement

In this method the fertilizer is placed in bands. These bands could be continuous or discontinuous (close to the seed or transplanted plant).As such band placement is of two types:

Hill Placement

When the plants are spaced 1m or more on both sides fore example in cucurbits, fertilizers are placed close to the plant in bands on one or both sides of the plants. This is known as hill placement. The length and depth of the band and its distance from plant varies with the crop and the amount of fertilizer.

Row Placement

When the seeds or plants are sown close together in a row, the fertilizer is put in continuous bands on one or both sides of the row by hand or a seed drill. This method of application is known as row placement and can be practiced in potato and other vegetable crops.

Side Dressing

In this method, fertilizers are spread in between the rows or around the plants. As such, side dressing is a very broad term covering various practices of applying fertilizers. Application of nitrogenous fertilizers in between the rows by hand to broad-row vegetable crops such as cauliflower, cabbage tomato, brinjal and potato, will supply additional doses of nitrogen to the growing crop. Localized placement is more important with phosphorus and potassium than with nitrogenous fertilizers.

It would be desirable to add N as close as possible to the peak requirement of the crop. Split application of N improves N efficiency during vegetative growth stage of the crop. Placement of nitrogenous fertilizers at the time of sowing in medium-textured soil reduces volatilization losses.While potassium is commonly applied before or at planting. This time of application is more effective than side dressing, for opportunity is provided to incorporate the element in soil.

Phosphorus generally move only short distances from their points of placement. To make them available to the plants, phosphorus must be placed in the zone of root development for being utilized by the plant. Surface application after a crop is planted, not being in the zone of root activity, is of little value to row crops in the year of application. The placement of water- soluble phosphorus in bands tends to reduce contact with the soil and results in lesser fixation than in broadcast application. In contrast to phosphorus, the nitrate salts are mobile and move vertically or horizontally within the soil as the water moves.

Bibliography

Abbott, J. D., M. M. Peet., D. H. Willits., D. C. Sanders and R. E. Gough. 1986. Effects of irrigation frequency and scheduling on fruit production and radial fruit cracking in green house tomatoes in soil beds and in soil-less medium in bags. Scientia Hort. 28: 209-217.

Ackley, W. B and W. H. Krueger. 1980. Over-head irrigation water quality and cracking of sweet cherries. Hort. Sci. 15: 289-290.

Adams, P and L. C. Ho. 1992. The susceptibility of modern tomato cultivars to blossom-end rot in relation to salinity. J. Hort. Sci. 67: 827-839.

Adams, P. and L. C. Ho. 1993. Effects of environment on uptake and distribution of calcium in tomato and on the incidence of blossom end rot. Plant Soil. 154: 127-132

Adams, P. 1990. Effects of watering on yield, quality and composition of tomatoes grown in bags of peat. J. Hort. Sci. 65: 667-674.

Akeley, R. V., G. V. C. Hough land, and A. E. Schark. 1962. Genetic differences in Potato tuber greening, Am. Potato J. 39: 409.

Anderson, E. M. 1946. Tip burn of lettuce. Effect of maturity, air and soil temperature and soil moisture tension. Cornell Bull. 829.

Arnon, D. I. 1959. Nature 184-10

Asahina, E. 1956. The freezing process of plant cell. Contrib. Inst. Low Temp. Sci. 10: 83-126.

Asahina, E. 1978. Freezing process and injury in plant cells. In: P. H. Li and A. Sakai{eds. }. Plant cold hardiness and freezing stress. Academic Press, New York. p. 17-36.

Asher, C. J, Edwards, D. G and Howeler, H. R. 1980. Nutritional disorders in case of cassava Dept. of Agri. Univ. Queens land, Australia.

Bakker, J. C. 1990. Effects of days and night humidity on yield and fruit quality of green house egg plant (*Solanum melongena* L.) Hort. Sci. 65: 747-753.

Bar Tal, A and E. Pressman. 1996. Root restriction and potassium and calcium solution concentrations affect dry matter production, cation uptake and blossom-end rot in green house tomato. J. Am. Soc. Hort. Sci. 121: 649-655.

Bartholomew, E. T. 1915. A pathological and physiological study of black heart of potato tubers. Centralbl. f. Bakt. u. Par; 11 Abt. 43: 609-638.

Bauerle, W. H and Short, T. H. 1977. American Vegetable Grower, Vol. 25. p. 68-70.

Bela, R. M., J. S. Fenlon and L. C. Ho. 1996. Salinity effects on the xylem vessels in tomato fruit among cultivars with different susceptibility to blossom end rot. J. Hort. Sci. 71: 173-179.

Belda, R. M and L. C. Ho. 1993. Salinity effects on the net work of vascular bundles during tomato fruit development. J. Hort. Sci. 68: 557-564.

Belvins, D. G. and Lukaszewski, K. M. 1998. Boron in plant structure and function. Annual. Rev. Plant Physiology. Plant Mol. Biol. 49: 481-500.

Benton, J and Jones, Jr. 1998. The micronutrient. In Plant Nutrient Manual, CRC Press Boca Boston London, New York, Washington

Beraha, L and W. F. Kwolek. 1975. Prevalence and extent of eight market disorders of Western grown head lettuce during 1973-1974 in the greater Chicago, Illinois area. Plant Dis. Rptr. 59: 1001-1004.

Bhandari, K. R. 1963. Lal Bagh. 8: 56.

Bienz, D. R. 1968. Proceeding of American Society of Horticulture Science. Vol. 93. p. 429-433

Blum, A. 1988. Plant breeding for stress environments. CRC-Press, Boca Raton, FL.

Booth, A. 1963. The role of growth substances in the development of stolons in J. D. Ivins and F. L. Milthorpe{Eds}. The growth of the potato. Butter worth, London, 999-113

Borkowski, J. 1975. Influence of new synthetic growth regulators RW3, RW13 and RW14 on the growth, yield and health of lettuce in the glasshouse. Acta Agrobot. 28:253-262.

Bose, T. K. and M. G. Some. 1986. Vegetable Crops in India, Naya Prokash, Calcutta, India.

Brandy, C. J. 1993. Biochemical and molecular approaches to fruit ripening an senescence. Proceedings of Australian Center for International Agri. Research. Canberra (Champ, B. R., Higley, E and Johnnson, G. I eds. }. 50: 198-204.

Brecht, P. E., A. H. Kader, and L. L. Morris. 1973. The effect of composition of the atmosphere and duration of exposure on brown stain of lettuce. Amer. Soc. Hort. Sci. 95: 536-538.

Buck, R. W. Jr. and V. Akeley. 1957. Effect of maturity, storage temperature and storage time on greening of potato tubers. Am Potato J. 44:56.

Burton, W. G. 1982. Post harvest Physiology of Food Crops. Longman, London.

Care, H. J. 1947. A study of certain factors affecting buttoning of cauliflower. Cornell Uni. Thesis.

Ceponis, M. J; F. M. Porter and J. Kaufman. 1970. Rusty – brown discolouration a serious market disorder of Western winter head lettuce. Hort. Sci. 5: 219-221.

Charles, G., Rossignol, L. and Rossignol, M. 1992. Environmental effects on potato plants *in vitro*. J. Plant Physiol. 139: 708-713.

Choudhary, B. 1967. Bulb crops. In: Vegetable Crops of India, National Book Trust, New Delhi. p. 95.

Christensen, J. V. 1972. Cracking in cherries, iv: physiological studies of the mechanism of cracking. Acta Agr. Scand. 22:153-162. Coakley, S. M., R. N. Campbell and K. A. Kimble. 1973. Internal rib necrosis and rusty brown discolouration of Climax lettuce induced by lettuce mosaic virus. Phyto-pathology. 63: 1191-1197.

Collier, G. F. and V. C. Huntington. 1978. Physiological aspects of lettuce tip burn. Environmental conditions. Rpt. Natl. Veg. Res. Sta., Wellesbourne, 1977 p. 38-39.

Collier, G. F., Wurr, D. C. E. and Hungtington, V. C. 1978. The effect of calcium nutrition on the incidence of internal rust spot in potato. J. Agri. Sci. Camb. 91, 241.

Collier, G. F., Wurr, D. C. E and Huntington, V. C. 1980. The susceptibility of potato varieties to internal rust spot. J. Agri. Sci. Camb., 94, 407.

Cosidine, J. A., J. F. Williams and K. G. Brown. 1974. A model of studies on stress in dermal tissues of mature fruit of *vitis vinifera*: Criteria for producing fruit resistant to cracking. p. 611-617. In R. L. Bieleski *et al.* (eds.), Mechanism of regulation of plant growth. Bul. 12. Royal Society of Newzealand, Wellington.

Considine, J and K. Brown. 1981. Physical aspects of fruit growth: theoretical analysis of distribution of surface growth forces in fruit in relation to cracking and splitting. Plant Physiology. 68: 371-376.

Corgan, J. N and D. J. Cotter. 1971. The effects of several chemical treatments on tip burn of head lettuce. Hort. Science. 6: 19-20.

Corns, J. B. 1937. A study of the influence of certain factors on the internal structure of the tomato fruit as related to puffing. Cornell Uni. Thesis.

Cotner, S. D., E. E. Burns and P. W. Leeper. 1969. Pericarp anatomy of crack resistant and susceptible tomato fruits. J. Am. Soc. Hort. Sci. 94: 136-137.

Cox, F. F. 1977. Pepper-spot in white cabbage-a literature review, A-DAS Q, Rev. 25-81.

Crew, H. J. 1947. A study of certain factors affecting Buttoning of cauliflower. Cornell Univ. Thesis.

Crisp, P., Collier, G. P and Thomas, T. H. 1976. The effect of boron on tip burn and auxin activity. Scientia Horticulture. 5, 215-226.

Cromack, H. . T. H. 1981. Purple discolouration of the flesh of potato tubers. DAS, Trawscoed.

Cutting, J. G. M., Wolstenholme, and B. N. Hardy, J. 1992. Increasing relative maturity alters the base mineral composition and phenolic concentration of avocado fruit. J. Horti. Sci. 67, 761-768.

Davis, W. B. 1926. Physiological investigation of the black heart of potato tuber. Bot. Gaz. 81: 323-338.

Dekock, P. C., Inkson, R. H and Hall, A. 1982. J. Plant Nutr. 5:57-62.

Dekreij, C., J. Panse., B. J Van Goor and J. D. J. Van Doesburg. 1992. The incidence of calcium oxalate crystals in fruit walls of tomato as affected by humidity, phosphate and calcium supply. J. Hort. Sci. 67: 45-50.

Den Outer, R. W. 1989. Internal browning of witloof chicory(*Cichorium intybus* L.). J. Hort. Sci. 64: 697-704.

Den Outer, R. W. and W. H. L. Van veenendaal. 1988. Gold speckles in tomato fruits (*Lycoperscion esculentum* Mill.). J. Hort Sci. 63: 645-649.

Dickey, R. D. 1977. Nutritional deficiencies of woody plants used in Florida Landscape. Fla. Agri. Expt. Sta. Press Bul. 791.

Dickey, R. D. and J. N. Joiner. 1966. Identifying deficiencies in foliage plants. Sou. Flor. & Nurseryman. 79(20): 38. p. 42-43.

Diehl, H. C. 1924. The chilling of tomatoes. U. S. Department Agr. Circ. 315: 1-6.

Dyson, P. W and Digby, J. 1975. Effect of calcium on sprout growth and subtropical necrosis in Majestic potatoes. Potato Res. 18: 290-305.

Eaton, F. M. 1942. Toxicity and accumulation of chloride and sulphate salts in plants. J. Agri. Res. 64: 357-399.

Ehret, D. L., T. Helmer and J. W. Hall. 1993. Cuticle cracking in tomato fruit. J. Hort. Sci. 68: 195-201.

Ellison, J. H., G. Vest and R. W. Langlois. 1981. Jersey centennial asparagus. Hort. Sci. 16 (2): 349.

Emsweller, S. l. 1932. An hereditary type of pithiness in celery. Proc. Am. Soc. Hort. Sci. 29: 480-85.

Evens, H. J. 1959. The bio-chemical role of iron in plant physiology, p. 89-110. Bul. 15, Duke Univ. School of Forestry, Durham, N. C

Fife, F. M and E. Carsner. 1945. The tip burn of sugarcan with special reference to some light and nitrogen relations. Phytopathology. 35: 910-920.

Filippov, L. A. 1961. Agrobiologica, p. 628 (Hort. Abstr. 32: 116).

Finney, E. E and Findlen, J. 1967. Influence of pre-harvest treatment upon turgor of Katahdin potatoes. Am. Potato J. 44. 383.

Finney, E. E and Morris, K. H. 1973. X-ray images of hollow heart potatoes in water. Am. Potato Jour. 32: 154-67.

Foster, A. C. 1937. Environmental factors influencing the development of blossom-end rot of tomatoes. Phytopathology. 27: 128-129.

Foy, C. D., Chaney, R. L. and White, M. C. 1978. The physiology of metal toxicity in plants. Annu. Rev. Pl. Physiol. 29: 511-566.

Frazier, W. A. 1934. A study of some factors associated with the occurrence of cracks in the tomato fruit. Proc. Am. Soc. Hort. Sci. 325: 519-523.

Frazier, W. A. 1935. A study of some factors associated with the occurrence of cracks in tomato fruit. Proc. Am Soc. Hort. Sci. 32: 519-523.

Frazier, W. A. 1947. A final report on studies of tomato fruit cracking. TGC Report. No. 1, p. 5.

Gajarathnam, S and Z. Bano. 1987. Pleurotus muse rooms, Part IB Pathology, *in vitro* and *in vivo* growth requirements and world status, CRC. Rev. Food Sci. Nutr. 26: 243.

Gallagher, P. A. 1972. Proc. Por. Glass house Conf. Durbin. p. 13-18.

Gerretson, F. C. 1950. Maganese in relation to photosynthesis. 11. Redox Potentials of illuminated crude chloroplast suspension. Pl. and Soil. 2 : 150-193.

Gerendas, J., Zhu-Zhujun., Bendixen, R., Ratcliffe, R. G., Sattelmacher, B. Zhu, Z. J. 1997. Physiological and biochemical processes related to ammonium toxicity in higher plants. Zeitschrift fur-Pflanzenernahrung-und-Bodendunde. 160:239-251.

Geraldson, C. M. 1954. The control of black heart of celery. Proc. Am. Soc. Hort. Sci. 63: 353

Gill, P. S and K. S Nandpuri. 1977. Comparative resistance to fruit cracking in tomato. Indian J. Agr. Sci. 40: 89-98

Goossens, H. 1988. Uitgangspunten Goudspikkels. Internal communication of the auction West led-Zuid.

Gray, D and Hughes, J. C. 1978. Tuber quality in the potato crop (ed. P. M Harris), Chapman and Hall. London, p. 504-44.

Greathead, A. S., D. Ririe and E. C. Maxie. 1966. A new disorder of straw berry fruit plant. Dis. Rep. 50: 327.

Grierson, D and A. A. Kader. 1986. Fruit ripening and quality. p. 241-280. In: J. G. Atherton and J. Rudich (eds). The tomato crop: A scientific basis for improvement. Chapman and Hall, London.

Gull, D. D and F. M. Isengerg. 1958. Light and off flavour development in potato tuber exposed to flurescent light. Proc. Am. Soc. Hort. Sci. 71:446.

Hanger. B. C. 1979. The movement of calcium in plants. Comm. Soil Sci and Pl. Anal. 10: 171-193.

Hankinson, B and V. N. M. Rao. 1979. Histological and physical behaviour of tomato skins susceptible to cracking. J. Am. Soc. Hort. Sci. 104: 577-581.

Hansen, E. 1961. Climate in relation to post harvest physiological disorders of apples and pears. Proc. Oregon Hort. Soc. 53: 54-58.

Hartung, A. C., Putnam, A. R and Stephens, C. T. 1989. Hort. Sci. Vol. 114. p. 144-148.

Harvey, H. B and R. C. Wright. 1922. Frost injury to tomatoes. U. S. Dept. Agr. Bull. 1099.

Hassen, M. S. 1978. Effects of nitrogen fertilization and plant density on yield and quality of tomatoes in Sudan Gezira. Acta Hort. 84: 79-84.

Hurd, R. G. 1978. The root and its environment in the nutrient film technique of water culture. Acta. Hort. 82: 87-97.

Hurd, R. G. and C. J. Graves. 1983. Interaction between air and root temperature effects on tomatoes in nutrient film culture. Annu. Rpt. Glasshouse Crops Res. Inst.

Hayman, G. 1978. The hair like cracking of last season. Grower. 107: 3-5.

Hayslip, N. C and Lley, J. R. 1967. Proc. Fla. Sta. Hort. Sci. 79: 132-9.

Heald, D. F. 2006. Encyclopedia of plant diseases in Agriculture and Horticulture. Asiatic Publishing House, Delhi, India.

Hewitt, E. J and T. Takagi. 1975. Plant mineral nutrition. John Wiley and Sons, New-York.

Hiller, L. K and Koller, D. C 1984. Effect of early season soil moisture levels and growth regulator applications on internal quality of Russet Burbank Potatoes. Proc. Wash. State Potato Conf, . 23. 67.

Hiller, L. K., Koller, D. C and Thornton, R. E. 1985. Physiological disorders of potato tubers, in potato physiology{ ed. P. H. Li}. Academic Press, London. p. 389-455.

Hiller, L. K., Koller, D. C and Van Den Burgh, R. W. 1979. Brown center of potatos: What have we learned ? Proc. Wash. State Potato Conf. 18, 21.

Hindawi, I. J. 1970. Air pollution injury to vegetation. U. S. Dept. Health, Education and Welfare, Washington, D. C.

Hobson, G. E. 1987. Low temperature injury and the storage of ripening tomatoes. J. Hort. Sci. 62: 55-62.

Hogetop, K. 1930. Untersuchungen iiber de Einfluss der temperature auf Keimung und Labensdauer der Kar Toffel Kuolle BA. Archir. 30, 350-413.

Hogge, M. C. 1989. Effects of site, season and husbandry on yield and processing quality of potato variety Pentland Dell, Ph. D. Thesis, University of Cambridge. P 208.

Holmes, F. O 1949. Candian Journal of Plant Science. Vol. 54. p. 58-59.

Hooker, W. J. (ed). 1981. Compendium of potato Diseases. Amer. Phytopath. Soc St. Paul. MN, USA. p. 125.

Horsfall, J. G. 1948. An unusual occurance of tomato blossom end rot. Plant Disease Rptr. 32: 351.

Ho, L. C. 1999. The physiological basis for improving tomato fruit quality. Acta. Hortic. 487: 33-40.

Ho. L. C., R. Belden., M. Brown., J. Andrews and P. Adams. 1993. Uptake and transport of calcium and the possible cause of blossom-end rot in tomato. J. Expt. Bot. 44: 509-518.

Hurd, R. G and C. J. Graves. 1983. Interaction between air and root temperature effects on tomatoes in nutrient film culture. Annu. Rept Glasshouse Crops Res. Inst.

Igbokwe, P. E., S. C. Tiwari., J. B. Collins and L. C Russels. 1978. Tomato cultivars response to foliar calcium and magnesium. Applications applications. J. Mississippi Acad. Sci. 32:123-131.

Iritani, W. M. 1981. Growth and pre harvest stress and processing quality of potatoes. Am. Potato J. 58, 71.

Jackman, R. L and D. W. Stanley. 1992a. Area and perimeter-dependent properties and failur of mature green and red ripe tomatoe pericarp tissue. J. Text Stud. 23: 461-505.

Jackson, T. L., Mc Bride, R., Powelson, M. L. *et al*. 1984. Soil fertility, plant nutrition and plant disease interactions affecting potatoes. Oregon State Uni. Agric. Exp. Stn. Spec. Rept. July. p. 34-37.

Jacobson, J. S and Hill, A. C. 1970. Recognition of Air Pollution Injury to Vegetion. A pictorial Atals. Air Pollution Control Association, Pitsburg.

Jefferies, R. A and Mac Kerron, D. K. L. 1987. Observations on the incidence of tuber growth cracking in relation to weather patterns. Potato Res. 30, 613.

Jenkins, J. E., Wiggell, D and Fletcher, J. T. 1962. Progress Export Exp. Hus. Fms. And Exp Hort. States. p. 40-41.

Joiner, J. N and W. E. Waters. 1970. The influence of cultural conditions on the chemical composition of six tropical foliage plants. Proc. Trop. Res. Am. Soc. Hort. Sci. 14. p. 254-267.

Johnson, H. Jr., D. R. Woodruff and T. W. Whitaker. 1970. Internal rib necrosis of head lettuce in Imperial Valley. Calif. Agr. 24 (9): 11-11.

Jolivette, J. P and J. C Walker. 1943. Effect of boron deficiency on the history of garden beet and cabbage. J. Agri. Res. 66: 167-182.

Jones, Jr. J. B., B. Wolf and H. A. Mills. 1991. Plant Analysis Hand book. Macro Publishing Inc. Georgia, USA., p. 213. J. Rudich(ed). The tomato Crop: A scientific basis for improvement. Chapman and Hall London.

Kader, A. A. 1992. Post harvest biology and technology: an overview, p. 20. in: Kader, A. A{eds. }, Post harvest of horticultural products. 2nd ed. Univ. California, Division of Agriculture and Natural resources, Publication 3311.

Kader, A. A., Brecht, P. E., Woodruff, R, and Morris, L. L. 1973. Influence of carbon monoxide, carbon Dioxide and oxygen levels on brown stain, respiration rate, and visual quality of lettuce. Journal of American Society for Horticultural Science 98, 485-488.

Kader, A. A., P. E. Brecht., R. Woodroff and L. L. Morris. 1973. Influence of carbon monoxide, CO2 and O2 levels on brown stain, respiration rate and visual quality of lettuce. J. Amer. Soc. Hort. Sci. 98: 485-488.

Kallio, A. 1960. Effect of fertility level on the incidence of hollow heart. Am. Potato J. 37, 338.

Ke, D. Y. and Saltveit, M. E., Jr. 1986. Effect of calcium and auxin on russet spotting and phenylalanine ammonia-lyase activity in iceberg lettuce. Hort Science 21: p. 1169-1171.

Kidd, F and West. C. 1923. Brownheart a functional disease of Apple and pears. Spec. Rep. Fd. Invest Bd. D. S. I. R Lond. -12.

Kirkby, E. A. 1979. Maximizing calcium uptake by plants. Comm. Soil Sci. and Plant Anal. 10: 89-115

Koske, T. J., J. E. Pallas Jr. and J. B. Jones Jr. 1980. Influence of ground bed heating and cultivar on tomato fruit cracking. Hort Sci. 15: 760-762.

Kramer, P. J. 1980. Drought stress and the origin of adaptations. p. 20. In, N. C. Turner and P. J. Kramer{eds. }, Adaptation of plants to water and high temperature stress. Wiley, Toronto.

Krijthe, N. 1963. Observation on the sprouting of seed potatoes. Eur. Potato Jour. 5. 316-33.

Kruger, N. S. 1966. Tip burn of lettuce in relation to calcium nutrition. Queens land J. Agr. and Animal Sci. 23: 379-385.

Lampe, K. 1960. Die Widerstandsfahigkeit von Kartoffelknollen gegan beschadigungen. Eur. Potato J. 3. p. 13-29.

Lau, O. L. and S. F. Yang. 1975. Interaction of kinetin and calcium in relation to their effect on stimulation of ethylene production. Plant Physiology. 55: 738-740.

Lee, S. H and R. L. Cardus, 1949. Effect of certain growth regulating substances. Michingan Agri. Exp. Sta. Tech. Bull. 216.

Leone, I. A. 1977. Effects of atmospheric pollution on vegetation. Proc. International Symp. On Environmental Pollution and Toxicology. {Ed. Raychaudhari, S. P. and Gupta, D. S}. Today and Tomorrows Printers and Publishers. New Delhi. p. 1-9.

Levitt, J. 1980. Responses of plants to environmental stress. Vol. 1, 2nd ed. Academic Press New York.

Lewis, W. C and R. G. Rowbery. 1973. Some effects of depth of planting, time and height of hilling on 'Kennebec' and 'Sebago' Potatoes. Amer. Jour. 50: 301.

Lieten, F., Marcelle, R. D. 1993. Relationships between fruit mineral content and albinism disorder in strawberry. Ann. Appl. Biol. 123, 4433-439.

Lipton, W. J. 1963. Influence of max. air temperatures during growth on the occurrence of russet spotting in head lettuce. Proc. Amer. Soc. Hort Sci. 83: 590-595.

Lipton, W. J., J. K. Stewart and T. W. Whitaker 1972. An illustrated guide to the identification of some market disorders of head lettuce. USDA Marketing Res. Rpt. 950.

Lipton, W. J. 1977. Towards an explanation of disorders of vegetables induced by high Co2 or low O2. Proc, Second. Natl. CA. Res. Conf. Mich. State Univ. Hort. Rpt. 28: 137-141.

Lorenz, O. A. 1942. Internal break down of table beets N. Y{Cornell} Agri. Expt. Sta. Mem. 246.

Lutman, B. F. 1919. Tip burn of the potato and other plants. Vt. Agr. Exp. Sta. Bul. 214: 1-28.

Lyons, J. M. 1973. Chilling injury in plants. Annu. Rev. Physiol. 24: 445-446.

Mac Kerron, D. K. L. and Jefferies, R. A. 1986. The influence of early soil moisture stress on tuber snumbers in potato. Potato Res. 29, 299-312.

Macmillan, H. G. 1920. A frost injury of potatoes. Phytopath. 10: 423-424.

Macmillan, H. G. 1923. The cause of sun scald of beans. Photopath. 13: 376-380.

Manning, W. J., Godzik, B. and Musselman, R. 2002. Proceedings of the Ist. International Symposium on Labiatae. Advances in Production, Biotechnology and Utilization{ed. Cervelli, Ruffoni and Dalla Guda}. Acta Horticulture. No. 723: p. 178-183.

Marangoni, A. G., R. L. Jackson, and D. W. Stanely. 1995. Chilling associated softening of tomato fruit is related to increased pectinmethylesterase activity. J. Food. Sci. 60: 1277-1281.

Marcelle, R. D. 1984. Mineral analysis and storage properties of fruit. In: Martin-Prevel, P{Ed}. VIth international colloquiun for the optimization of plant nutrition. Vol. 2 A IONP/ GERDAT, Montppellier, p. 365-371.

Marlatt, R. B and J. K. Stewart. 1956. Pink rib of head lettuce. Plant Disease Reporter 40, 742-743.

Marschner and Ossenberg-Neuhaus 1977. Effect of 2, 3, 5-triiodobenzoic acid (TIBA) on calcium translocation and catio exchange capacity in sunflower. Z. Pflanzenphysiol. Bd. 85: 29-44.

Marinos, N. G. 1962. Studies on submicroscopic aspects of mineral deficiencies. 1. Calcium Deficiency in the stem apex of barley. Amer. J. Bot. 49: 834-841.

Marschner, H. 1985. Mineral nutrition of higher plants. 2nd edition. Academic Press, London, U. K.

Maynard, D. N. 1979 Ann. Review. J. Pl. Nut., 1. 1: 1-23.

Maynard, D. M., B. Gersten., E. F. Vlach and H. F. Vernell. 1963. The influence of plant maturity and calcium level on the occurrence of carrot cavity spot. Proc. Am. Soc. Hort Sci. 83: 506.

Maynard, D. M. and A. O. Lorenz. 1979. Controlled release fertilizers for horticultural crops. Hort. Rev. 1: 79-140.

Mengel, K and E. A. Kirkby. 1987. Principal of plant nutrition, Forth Edition. International Potash Institute, A. G. Berne, Switzerland.

Milad, R. E. and K. A. Shackel. 1992. Water relations of fruit end cracking in French prune. Am. Soc. Hort. Sci. 117: 824-828.

Miller, H. G., Ikawara, M and Pierce, L. C. 1991. Hort. Science Vol. 26. p. 1825-1527.

Misaghi, I. J., Matyac, C. A and Grogan, R. G. 1981b. Soil and foliar application of calcium chloride and calcium nitrate to control tip burn of head lettuce. Plant Disease 65, 821-822.

Misaghi, I. J and R. G. Grogan. 1978. Physiology bases for tip burn development in head lettuce. Phytopathology 68: 1744-53.

Moline, H. E and Lipton, W. J. 1987. Market diseases of beet, chicory, endive, escarole, globe artichoke, lettuce, rhubarb, spinach and sweetpotasoes. Agriculture Handbook Number 155, US Department of Agriculture Research Service, Washington, DC, USA. P. 86.

Morris, L. L and A. A. Kader. 1977. Physiological disorders of certain vegetables in relation to modified atmosphere. Proc. Second Natl. Ca Res. Conf. Mich. State Univ. Hort. Rpt. 28: 142-148.

Morris, L. L; A. A. Kader; J. A. Klaustermey and C. C. Cheyney. 1978. Avoiding ethylene concentration in harvest lettuce. Calif. Agr. 32. 6: 14-15.

Mulder, E. G. 1954. Molybdenum in relation to growth of higher plants and micro-organisms. Plant Soil. 5: 368-415.

Nelson, D. C and Thoreson, M. C. 1986. Relationships between tuber size and time of harvest to allow heart initiation in dry land Norgold Russet Potatoes. Am. Potato J. 63: 155.

Newhall, A. G. 1929. Studies on tip burn of head lettuce, Cornell Uni. Thesis.

Nonami, H., T. Fukuyama., M. Yamamoto., L. Yang., Y. Hashimoto., T. Ito., F. Tongnomi., T. Namiki., A. Nukaya and T. Maruo. 1995. Blossom-end rot of tomato plants may not be directly caused by calcium deficiency. Acta Hort. 396: 107-114.

Ohta, K., T. Hosoki and K. Inaba. 1995. The initiation of minute cuts and the occurrence of cracking in cherry tomato fruits. Acta Hort. 396: 251-256.

Olson, K. C., T. W. Tibbitts and B. E. Struckmeyer. 1967. Morphology and significance of laticifer rupture in lettuce tip burn. Proc. Amer. Soc. Hort. Sci. 91: 377-385.

Oursel, A., A. Lamant, L. Salsac, and P. Mazliak. 1973. Etude comparee des lipides et de la fixation passive due calcium dans less raciness et les fractions subcellulaires du *lupinus luteus* et de la *Vici faba*. Phytochemistry 12: 1865-1874.

Ozbun, J. L., Bontomel, C. E., Adik, S and Minges, P. A. 1967. Tomato fruit ripening. 1. Effects of potassium nutrition on occurrence of white tissue. Proc. Am. Soc. Hort. Sci. 91: 566-572.

Palevitch, D. and Kedar, N. 1963. Research Rep. Science and Agriculture, Hebrew University Jerusalem. p. 436 –537.

Palskill, D. A., Tibbits, T. W., and Williams, P. H. 1976. Enhancement of calcium transport to inner leaves of cabbage for prevention of tip-burn. J. Am. Soc. Hort. Sci. 97: 397-4028.

Palzkill, D. A. and Tibbitts, T. W. 1977. Evidence that root pressure flow is required for calcium transport of head leaves of cabbage, Plant Physiol. 60: 854.

Palzkill, D. A., Tibbits, T. W and Williams, P. H. 1976. Enhancement of calcium transport to inner leaves of cabbage for prevention of tip burn. J. Am. Soc. Hort. Sci. 101: 645.

Pandey, U. G. and J. Singh. 1993. Agro-techniques for onion and garlic, in Advances in Horticulture. Vol 5, Vegetable Crops (K. L. Chadha and G. Kalloo, eds} Malhotra Pub. House. New Delhi. p. 433.

Pandey, U. C., J. Singh and V. Virender. 1990. Onion and Garlic as cash crop: Problems and Solutions. Haryana Farming, 19: 8.

Pantastico, C. B. 1975. Post harvest Physiology, Handling and Utilization of Tropical and sub-tropical fruits and vegetables. AVI Publishing West Post Conn. Hort. Reviews. Vol. 4. 1982, p. 260.

Parsons, C. S and R. H. Day. 1970. Freezing injury of root crops: Beets, carrots, parsnip, radishes And turnips. US Dept. Agric. Marketing Res. Rep. P. 866

Pasture, E. 1971. Tuinbouwwberichten Belgium. 35: 328-30.

Peck, N. H., M. H. Dikson and G. E. Mac Donald. 1983. Tip burn susceptibility in semi-isogenic inbred lines of cabbage influenced by nitrogen. Hort. Science. 18:726.

Peet, M. M. 1992. Fruit cracking in tomato. Hort. Technology. 2: 216-219, 222-223.

Peet, M. M and D. H. Willits. 1995. Role of excess water in tomato fruit cracking. Hort. Sci. 30: 65-68.

Pock, N. H., M. N. Dickson and G. E. Mac Don land. 1983. Tip burn susceptibility in semi-isogenic inbred lines of cabbage influenced by nitrogen. Hort. Sci. 18: 726.

Purvis, C. R and W. J. Hanna. 1940. Vegetable crops affected by boron deficiency in Eastern Virginia, Va. Truck Exp. Sta. Bull. 105.

Purvis, E. R and R. W. Ruprecht. 1937. Cracked stem of celery caused by a boron deficiency in the soil. Fla. Bull. 307.

Raison, J. K. 1974. A biochemical explanation of low temperature. Stress in tropical and subtropical plants. R. L. Bieleski, A. R. Ferguson and M. M. Cresswell {ed}Mechanisms of regulation of plant growth. Roy. Soc. New Zealand, Bull. 12. Wellington, N. Z. p487-97.

Raleigh, G. J *et al*. 1941. Studies on the control of internal breakdown of table beets by the use of boron. N. Y. (Cornell) Agr. Ext. Sta. Bull. 752.

Raleigh, S. M and J. A. Chucka. 1944. Effect of nutrient ration and concentration on growth and composition of tomato plants and on the occurrence of blossom end rot of the fruit: Plant Physiology. 19: 671-678.

Ralph, R. K., S. Bullivant, and S. J Wojcik. 1976. Effect of kinetin on phosphorylation of leaf membrane proteins. Biochem. Biophys. Acta 421: 319-327.

Rao, R. R. 1966. Studies on the environmental factors controlling tip burn of lettuce. Ph. D. Dissertation, Univ. Of Wisconsin, Madison.

Ray chudhury, S. P and Lele, V. C. 1966. Indian Hort. 10: 39-43.

Rayle, D. L. and R. Cleland. 1977. Control of plant cell enlargement by hydrogen ions. Curr. Top. Dev. Biol. 11: 187-214.

Read, M. 1972. Growth and tip burn of lettuce: CO2 enrichment at different light intensity and humidity levels and rate of incorporation of carbon-14 assimilation into the latex. Ph. D. Dissertation, Univ. of Wisconsin, Madison.

Reynard, George, B. 1951. Inherited resistance to radical cracks in tomato fruits. Amer. Soc. Hort. Sci. Proc. 58: 231-244.

Richardson, S. J. 1982. Crop nutrition in nutrient film culture. Fert. Soc. London. Dec. 1981.

Richardson, L. T and Phillip, W. R. 1949. Low temperature breakdown of potatoes in storage. Scient. Agric. 29: 149-66.

Robbins, W. R. 1939. Relation of nutrient salt concentration to growth of the tomato and to the incidence of blossom end rot of the fruit. Ibid; 12-50.

Robbins, W. R., G. T. Nightingale and G. Shermerhorn. 1931. Premature heading of Cauliflower as associated with chemical composition of the plant. N. J. Bull. 509.

Rood, R. 1956. Relation of ethylene and post harvest temperature to brown spot of lettuce. Proc. Amer. Soc. Hot. Sci. 68: 296-303.

Rose, P. 1992. Never mind the food, the packing tastes great. Food Manuf. Int. 9: 27, 29-30.

Ryall, A. L and W. J. Lipton. 1972. Handling, Transportation and Storage of Fruits and Vegetables, Vol. 1. Vegetables and Melons, AVI Pub. Co. Westport, CT.

Sadik, S and P. A. Minges. 1966. Symptoms and histology of tomato fruits affected by blotchy ripening. Proc. Am. Soc. Hort. Sci. 88: 532.

Salisbury, F. B and Ross, C. W. 1992. Plant physiology. 4th ed. Wadsworth Publishing Company. Belmount, California, USA.

Sarin, M and H. K. Saxena. 1965. Ind. J. Plant Physiology, 8:136-144.

Schoorl, D and Holt, J. E. 1983. Cracking in potato. J. Texture Studies. 14, 61.

Schroeder, W. T. 1949. The control of blossom end rot of tomato with emulsified Hydrocarbon sprays. Phytopathology. 39: 21-22.

Seaton, H. L and G. F. Gray. 1936. Histological study of tissue from green house tomatoes affected by blotchy ripening. Jour. Agri. Res. 52: 217-224.

Sharma, R. R and Sharma, V. P. 2003. Plant growth and albisism disorder in different straw berry cultivars under Delhi conditions. Ind. J. Hortic. 61. 92-93.

Sharma, R. R and V. P. Sharma. 204. Plant growth and Albinism disorder in different straw berry cultivars under Delhi conditions. Indian J. of Hort. 61{1}, March 2004. p. 92-93.

Shear, C. B. 1975. Calcium related disorders of fruits and vegetables. Hort. Sci. 10: 361-365.

Simon, E. W. 1978. The symptoms of calcium deficiency in plants. New Phytol. 80:1-15.

Singh, A. K., M. K. Singh., A. K. Srivastava and A. K. Singh. 2006. Chilling injury in fruits and vegetables and its remediation. Agricultural update, Vol. 1. 3. 8-10.

Singh, A. L., Y. C. Joshi., Vidya Choudhari and P. V. Zala. 1990. Effect of different sources of iron and sulphur on leaf chlorolysis nutrition uptake and yield of groundnut. Fertilizer Research. 24: 85-96.

Singh, J and B. S. Dankar. 1989. Effect of nitrogen, potash and zinc on growth, yield and quality of onion. Veg. Sci. 12:136.

Sorensen, J. N., A. S. Johansen and N. Poulsen. 1994. Influence of growth conditions of the value of crisped lettuce. 1. Marketable and nutritional quality as affected nitrogen supply, cultivar and plant age, Plant Foods Human Nutr. 46:1.

Spurr, A. R. 1959. Anatomical aspects of blossom end rot in the tomato with special reference to calcium nutrition, Hilgardia. 28:269.

Standberg, J. O., Darby, J. F., Walker, J. C and P. H Whilliams. 1969. Black peck a non-parasitic disease of cabbage, Phytopathology. 59: 1879.

Stewart, F. C and Mix, A. J. 1917. Black heart and the aeration of potato's in storage. Bull. N. Y. St. Agric. Exp. Stn. 436, 312-62.

Stewart, J. K., M. L. Ceponis and L. Beraha. 1970. Modified atmosphere effects on the market quality of lettuce shipped by rail. USDA. Mktg. Res. Rpt. 863.

Taverna, G. 1968. Ann Fac Sci Agrar Wapoli, Vol. 3 p. 299-306.

Thibodeau, P. O. and P. L. Minotti. 1969. The influence of calcium on the development of lettuce tip burn. J. Amer. Soc. Hort. Sci. 94: 372-376.

Thomas, Florence. 1948. Studies of cracking of tomato fruits with emphasis on method of selecting resistant plants from segregating progenies. Cornell Univ. Thesis.

Thompson, H. C. 1927. Experimental studies of cultivation of certain vegetable crops. Cornell Memoir 107.

Thoms, R. J and K. Pearson. 1995. The heat sensitive stage of Broccoli flower development. Meeting of America Society for Horticultural Sciences Montreal, Quabec.

Tibbitts, T. W. and M. Read. 1976. Rate of metabolite accumulation into latex of lettuce and proposed association with tip burn injury. J. Amer. Soc. Hort. Sci. 101: 406-409.

Tibbitts, T. W. and R. R. Rao. 1968. Light intensity and duration in the development of lettuce tip burn. Proc. Amer. Soc. Hort. Sci. 93: 454-461.

Tibbitts, T. W., B. E. Struckmeyer and R. R. Rao. 1965. Tip burn of lettuce as related to release of latex. Pros. Amer. Soc. Hort. Sci. 86: 462-467.

Timoshenko, S and S. Wionowsky-Krieger. 1959. Theory of Plates and Shells. 2nd. Eds. Mc Graw-Hill, New-Yark.

Tisdal, S. L and W. L. Nelson. 1975. Soil fertility and fertilizers. Mac Millan, New –York.

Treshow, M. 1955 The etiology, development and control of tomato fruit tumor. Phytopathology. 45:132

Trudel, M. J. and Ozbun, J. L1970. J. Expt. Bot. 69: 881-6.

Tzeng, K. C., Kelman, A., Simmons, K. E. and Kelling, K. A. 1986. Relationship of calcium nutrient to internal brown spot of potato tubers and sub-apical tip necrosis of sprouts. Am. Potato J. 63, 87.

Van der Zaag, D. E and Meijers, C. P. 1970. black-spot-practical aspects. Proc. Trienn Conf. Eur. Assoc. Potato Res. 1969, 4, 93-103.

Vander Zaag, K. and Ffrench. S. 1987. Preliminary evaluations of foliar calcium applications with respect to yields and processing quality of potato cultivars. Atlantic and Norchip Ont. Potato Cultivar Eval. Assoc. Bull No. 2. p. 3-5.

Ventor, F. 1970. Angew Bot. Vol. 44, p. 263-270.

Verner, L. 1935. A physiological study of cracking in stayman wines ap apples: J. Agr. Res. 51: 191-222.

Verner, L. and E. C. Blodgett. 1931. Physiological studies of the cracking of sweet cherry: a preliminary report. Idaho Agr. Exp. Sta. Bull. 184.

Wadleigh, C. H. 949. Soil Science. 61: 225-38.

Walker, J. C. 1939. Internal black spot of garden beet. Physiology, 29: 120-128.

Walker, J. C. 1939. Phytopathology, Vol. 29. p. 120-128.

Walker, J. C., Whilliams, P. H and G. S. Pound. 1965. Internal tip burn of cabbage: Its control through breeding, Univ. Wisc. Res. Bull. 258.

Walker, J. C. 1943. Boron deficiency in garden beet and sugar beet. Jour. Agri. Res. 66: 97-123.

Walker, R. B. 1948. Molybdenum deficiency in serpentine barren soils. Science, 108: 473-475.

Wallace, T., Croxall, H. E and Pickford, P. T. H. 1942. Annual Report, Long Aston Agriculture and Horticulture Research Station Vol. 1, p. 25-31.

Walter, T. E. 1967. Russetting and cracking in apples: a review of world literature. Rep. East Mall. Res. Sta. Forfor 19966. p. 83-95.

Ward, G. M. 1973. Cannadian Journal of Plant Sciences Vol. 53. p. 169-174.

Waring, E. j., R. D. Wilson and N. S. Shirlow. 1949. Whiptail of cauliflower. Agri. Gaz. N. S. Wales, 60: 21-26.

Whipple, O. C. 1941. Injury to tomatoes by chilling. Phytopathology. 31: 1017-1022.

Whiteman, T. M. 1957. Freezing points of fruits, vegetables and florist. Stocks. U. S. Dept. 12313B(Rev.) Nat. Canners Assoc. Res. Lab. San-Francisco.

Wiebe, H. J. 1967. Investigations of tip burn on lettuce. Gartenbauwissen-schaft 32: 375-385.

Wiersum. L. K. 1966. Calcium content of fruits and storage tissues in relation to the mode of water supply. Acta Bot. Neerl. 15: 406-418.

Wilcox, G. E. andPfeiffer, C. L. 1990. Temperature effect on seed germination, seedling root development and growth of several vegetables. J. Plant Nut. 13: 1393-1403.

Wills, R. H., T. H. Lee., D. graham, W. B. Mc glasson and E. G. Hall. 1981. Post harvest: an introduction to the physiology and handling of fruit and vegetables. Avi. Pub. Co. West pot. CT.

Willumsen, J., K. K. Peterson and K. Kaach. 1996. Yielded and blossom-end rot of tomato as affected by salinity and cation activity ratio's in root zone. J. Hort. Sci. 71: 81-98.

Winosor, G. W and P. Adams. 1976. Changes in the composition and quality of tomato fruit throughout the season. Annu. Rep. Glasshouse Crops. Res. Inst. P. 134-142.

Winsor, G. W., J. N. Davis and M. I. E. Long. 1961. Liquid feeding of glass house Tomatoes: the effects of potash concentration on fruit quality and yield. Jour. Hort. Sci. 36: 254

Woods, M. J. 1964. Ir. Journal of Agriculture of Research, Vol. 3. p. 17 –27.

Wright, R. C. and Diehl, H. C. 1972. Freezing injury to potato. Tech. Bull. US. Dep. Agri. 27. p. 23

Wright, R. C. *et al*. 1931. Effect of various temperatures on storage and ripening of tomatoes. U. S. Dept. Agr. Tech. Bul. 268: 1-34

Yarnell, S. H *et al*. 1937. Factors affecting the amount of puffing in tomatoes. Tex. Agri. Expt. Sta. Bul. 541

Young, P. A. 1946. Texas Agriculture Experiment Station Circular. p. 113.

Young, P. A. 1947. Cuticle cracks in tomato fruits. Phytopathology. 37: 143-145.

Index